SCIENCE
A CLOSER LOOK

BUILDING SKILLS

Reading and Writing Workbook

McGraw Hill Education

Contents

Contents

Contents

Contents

Contents

Contents

Insect-Eating Plants

**Read the Unit Literature pages in
your textbook.**

✏️ **Write About It**

Response to Literature

1. What is the main idea? How do you know?

_____

– – – – – – – – – – – – – – – – – – –

_____

– – – – – – – – – – – – – – – – – – –

_____

– – – – – – – – – – – – – – – – – – –

2. How do the captions tell about the main idea?

_____

– – – – – – – – – – – – – – – – – – –

_____

3. Draw a picture of one of the insect-eating

plants. Use a different sheet of paper.

Plants Are Living Things

Fill in the important ideas as you read the chapter. Use the words in the box.

air	grow	parts	water
change	leaves	roots	
flowers	make new things	living things	
food		stems	

What are plants?

Plants are _____
_____

Plants have _____
_____

What living things need

What living things do

Learning About Living Things

Use your textbook to help you fill in the blanks.

What are living and nonliving things?

1. Plants, animals, and people are

   _____

   _ _ _ _ _ _ _ _

   _____ things.

   _____

   _ _ _ _ _ _ _ _

2. All living things _____

   and change.

   _____

   _ _ _ _ _ _ _ _

3. They all need _____ ,

   air, and water to live.

   _____

   _ _ _ _ _ _ _ _

4. Living things can also make _____

   living things like themselves.

5. Things that do not grow or change are

   _____

   _ _ _ _ _ _ _ _

   _____ .

Name _____

Why are plants living things?

_____

6. Plants need air, _____ ,

and food, like other living things.

_____

_ _ _ _ _ _ _ _ _ _ _ _

7. Plants use air, water, and _____

to make their own food.

_____

_ _ _ _ _ _ _ _ _ _ _

8. Plants need nutrients and _____

to grow, too.

Critical Thinking

9. Compare a plant to a nonliving thing.
Tell how they are alike and different.

_____

_ _ _ _ _ _ _ _ _ _ _ _ _ _

_____

_____

_ _ _ _ _ _ _ _ _ _ _ _ _ _

_____

_____

_____

Learning About Living Things

Circle the best answer that tells about each picture.

I.

living   nonliving

2.

living   nonliving

3.

needs sunlight and nutrients

does not need sunlight and nutrients

4.

needs space to grow

does not need space to grow

Learning About Living Things

Fill in the blanks. Use the words from the box.

change	grow	living	nonliving

All living things need food and water
to live. Plants, animals, and people are

_____

_ _ _ _ _ _ _ _ _ _ _

_____ things. They can

_____

_ _ _ _ _ _ _ _ _ _ _

_____ and change.
They also need air and space to grow.

There are many other things that do
not need food, water or air. These are called

_____

_ _ _ _ _ _ _ _ _ _ _

_____ things. They do

_____

_ _ _ _ _ _ _ _ _ _ _

not grow or _____ .
Rocks, cars, and toys are nonliving things.

Parts of Plants

Use your textbook to help you fill in the blanks.

What are the parts of plants?

1. Plants have special _____Parts_____
to help them get what they need to live.

2. Leaves, _____roots stem_____ and roots
are some of these parts.

3. Plant parts look _____different_____
on different kinds of plants.

4. Plants also get what they need from the
_____space_____ around them.

Name _____

What do plant parts do?

5. Plant ___Leaves___ use sunlight
and air to make food.

6. Stems help carry food from the leaves to

other ___parts___ of the plant.

7. Plants take in ___water___ and
nutrients through their roots.

8. Roots keep the ___plants___ in
the ground.

Critical Thinking

9. Each part of a plant has a special job. Why?

_____

_____

_____

_____

Parts of Plants

Match each vocabulary word to the sentence that tells about it.

1. leaves

2. stems

3. roots

4. parts

a. These take in nutrients and water from the soil.

b. These help plants get what they need to grow.

c. These carry food and water to other plant parts.

d. These use sunlight and air to make food.

Parts of Plants

Fill in the blanks. Use the words from the box.

leaves	parts	~~roots~~

Most plants can not move around
to get food and water. They have

parts that help them get

what they need from where they live.

Plants use _leaves_

to make food. Plants use

root to get nutrients

and water from the soil. Plants use stems to

carry nutrients and water to leaves and the

rest of the plant.

© Macmillan/McGraw-Hill

Meet General Sherman

Write About It

Tell about a tall plant that you have seen. On a separate sheet of paper, draw it and label its parts.

Planning and Organizing

Use the web. Write the name of the plant in the center. Write describing words in the circles.

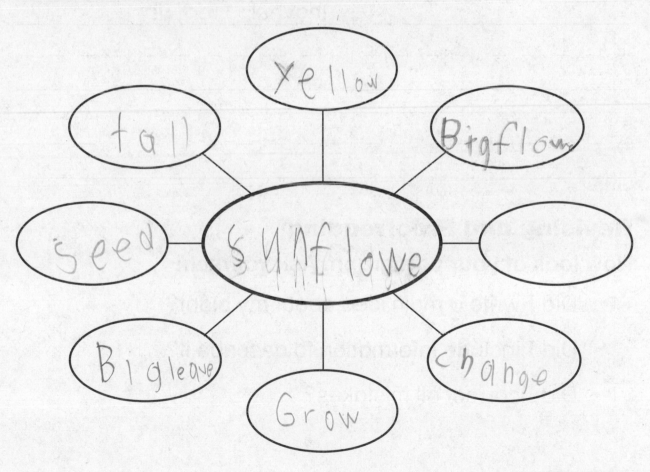

Name _____

Write about a tall plant. Start with a main idea. Write sentences that describe the plant.

_____

- - - - - - - - - - - - - - - - - - - -

_____

- - - - - - - - - - - - - - - - - - - -

_____

- - - - - - - - - - - - - - - - - - - -

_____

- - - - - - - - - - - - - - - - - - - -

_____

Revising and Proofreading

Now look at your paragraph. Ask yourself:

▶ Did I write a main idea about my plant?

▶ Did I include information to describe it?

▶ Did I correct all mistakes?

Name _____

Different Plants

Use your textbook to help you fill in the blanks.

How are plants different?

1. Plants can be ___fall___ or short.

2. They can have thick ___stems___
   or thin stems.

3. Trees are plants with thick stems called
   ___trunk___.

4. Some plants have colorful ___flowers___.

5. Plants can have different
   ___leaves___, too.

© Macmillan/McGraw-Hill

Name _____

Which plant parts can you eat?

6. You can eat the __*roots*__ ,
stems, and leaves of many plants.

7. Some foods, like lettuce, are
__*leaves*__ .

8. Others, like carrots, are __*roots*__ .

9. It is not safe to __*eat*__ all
plant parts.

Critical Thinking

10. Tell one way that plants can be different.
Tell another way they can be the same.

_____

_____

_____

_____

Different Plants

Match the words in the box to the correct statements on the left.

| leaves | roots | seeds | stems | trunks |

1. These are tree stems.

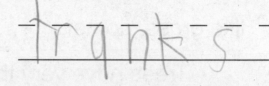

2. Carrots and potatoes are these plant parts that you can eat.

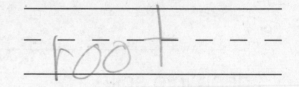

3. When you eat lettuce and spinach, you eat these.

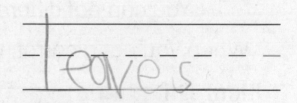

4. You can eat these parts of coconuts and sunflowers.

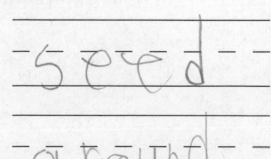

5. Grass has many of these thin things.

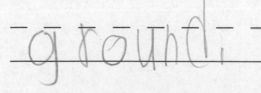

Name _____

Different Plants

Fill in the blanks. Use the words from the box.

root	trunks

Some plants, like trees, grow tall. Others, like grass, do not.

Trees have very thick stems called

trunks . Their roots grow

deep into soil.

You can eat different plant parts. When you eat a carrot, you eat the

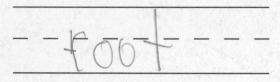 root of a plant.

Strawberry Fields

Read the Reading in Science pages in your book. Look for the main idea and details. Remember, the main idea is the most important idea. Details give more information about it.

Fill in the web below. Write the main idea in the top circle. Write three details in the other circles.

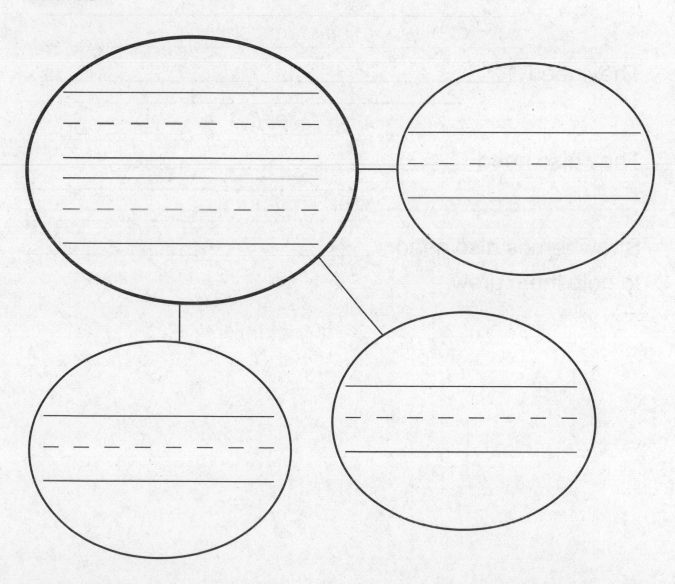

✏ Write About It

Find Main Idea and Details. What do strawberry plants need to grow? Finish this summary. Use the Main Idea and Details web you made on page 17.

Strawberry plants need three things to grow.

_____

_ _ _ _ _ _ _ _ _ _ _ _ _ _ _ _ _ _ _ _

They need _____ .

_____

_ _ _ _ _ _ _ _ _ _ _ _ _ _ _ _ _ _ _ _

They also need _____ .

_____

_ _ _ _ _ _ _ _ _ _ _ _ _ _ _ _ _ _ _ _

Strawberries also need _____
to help them grow.

Plants Are Living Things

Fill in the missing letters for each word.

1. Things that do not grow or

   change are called

   n __ __ __ l __ __ __ __ __ __ __ g things.

2. Plants, animals, and people are all

   __ __ iv __ __ n __ __ things.

3. All living things g __ __ __ __ w and change.

4. Nonliving things do not need food, air, or

   __ __ a __ __ er.

Name _____

**Write the word from the box that tells about
each picture. Then complete the sentence
below.**

living	nonliving	space

5.

_____

_ _ _ _ _ _ _ _ _ _ _ _ _

_____

6.

_____

_ _ _ _ _ _ _ _ _ _ _ _ _

_____

7.

_____

_ _ _ _ _ _ _ _ _ _ _ _ _

_____

8.

_____

_ _ _ _ _ _ _ _ _ _ _ _ _

_____

9. Living things need food, water, air, and

_____

_ _ _ _ _ _ _ _ _ _ _ _

_____ to grow.

Plants Grow and Change

Fill in the important ideas as you read the chapter. Use the words in the box.

adult	desert	parent	seedling	sprouts
Arctic	leaves	rainforest	seeds	

How Plants Grow and Change

How does a seed grow?

How can plants start?

seeds

seedling

sprouts

Where do plants grow?

Arctic

desert

rainforest

Flowers, Fruits, and Seeds

Use your book to help you fill in the blanks.

Why are flowers important?

_____

_ _ _ _ _ _ _ _ _

1. Some _____ grow flowers

   to help them live.

   _____

   _ _ _ _ _ _ _ _ _

2. A _____ is the part of a

   plant where seeds are made.

   _____

   _ _ _ _ _ _ _ _

3. A _____ is the plant part

   that can grow into a new plant.

   _____

   _ _ _ _ _ _ _ _ _

4. Some plants grow _____

   to protect the seeds.

   _____

   _ _ _ _ _ _ _ _ _

5. People and animals can _____

   fruits of many plants.

What are the parts of a seed?

_____

- - - - - - - - - - - -

6. Seeds can have different _____

and shapes.

_____

- - - - - - - - - - - -

7. All seeds need water, _____ ,

and warmth to grow.

_____

- - - - - - - - - - - -

8. Wind and _____ can

move seeds to new places.

Critical Thinking

9. Do you think a seed is a living thing?

Why or why not?

_____

- -

_____

- -

_____

Flowers, Fruits, and Seeds

Solve these plant riddles. Use the words in the box.

flower	fruit

1. I'm hard or I'm juicy.

I may be good to eat.

My job is to protect seeds.

At that, I can't be beat!

_____

– – – – – – – – – –

I am a _____ .

2. I'm bright and colorful.

I may smell good too.

My job is to make seeds.

I'm important!

It's true!

_____

– – – – – – – – – –

I am a _____ .

Flowers, Fruits, and Seeds

Fill in the blanks. Use the words from the box.

flowers	fruit	part	plant	seeds

Plants have different parts. Each plant

_____

_____ is important.

A plant's _____ make

seeds. These _____ can

grow into new plants. A plant's

_____ protects the seeds.

Inside each seed, a tiny

_____ grows.

How Plants Grow and Change

Use your book to help you fill in the blanks.

How do plants grow from seeds?

_____

_ _ _ _ _ _ _ _ _ _

1. A _____ is the way living

   things grow, live, and die.

   _____

   _ _ _ _ _ _ _ _ _ _

2. The life cycle of a _____

   can begin with a seed.

   _____

   _ _ _ _ _ _ _ _ _ _

3. A seed _____ into a

   young plant when it gets what it needs.

4. A young plant that comes from a

   _____

   _ _ _ _ _ _ _ _ _ _

   _____ is called a seedling.

   _____

   _ _ _ _ _ _ _ _ _ _

5. The seedling will grow into an _____

   plant.

How else do plants grow?

6. Not all plants _____ from

seeds.

7. A new plant may grow from a plant

_____

_____ that is cut.

8. New plants can also grow from

_____

_____ plants.

Critical Thinking

9. How are plant life cycles alike and different?

_____

_____

_____

_____

_____

_____

Name _____

How Plants Grow and Change

Complete the sentence that tells about each picture. Use the words in the box.

life cycle	seedling

_____

This _____ grows because it gets water and nutrients from the soil.

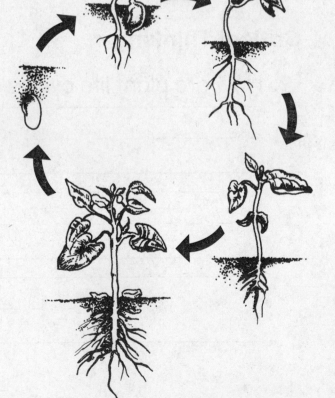

This picture shows the

_____

of a bean plant.

How Plants Grow and Change

Fill in the blanks. Use the words from the box.

adult	seedling	sprout	seeds

Plants grow in different ways. Some

_____

_ _ _ _ _ _ _ _ _ _

plants begin as _____

from a flower or fruit. A seed will

_____

_ _ _ _ _ _ _ _

_____ if it gets water

_____

_ _ _ _ _ _ _ _

and nutrients. A _____

is a young plant that comes up

from a seed. When it becomes an

_____

_ _ _ _ _ _ _ _

_____ plant, it makes

seeds of its own. Some plant life cycles do

not begin with seeds.

A Giant Grass

Read the Reading in Science pages in your book. Look for the most important ideas. Summarize them in the chart below. Remember, when you summarize, you retell the most important ideas.

Summary

_____
- -
_____
- -
_____

Idea 1	Idea 2	Idea 3
_____	_____	_____
_____	_____	_____
_____	_____	_____
_____	_____	_____
_____	_____	_____

1. How is bamboo like other plants you have seen? How is it different?

_____
_ _ _ _ _ _ _ _ _ _ _ _ _ _ _
_____
_____
_ _ _ _ _ _ _ _ _ _ _ _ _ _ _
_____
_____
_ _ _ _ _ _ _ _ _ _ _ _ _ _ _
_____

Write About It

How do you know bamboo makes life easier for people? Use the chart you made to help you write your answer.

_____
_ _ _ _ _ _ _ _ _ _ _ _ _ _ _
_____
_____
_ _ _ _ _ _ _ _ _ _ _ _ _ _ _
_____
_____
_ _ _ _ _ _ _ _ _ _ _ _ _ _ _
_____

Plants Live in Many Places

Use your book to help you fill in the blanks.

Where do plants live?

_____
_ _ _ _ _ _ _ _ _ _

1. You can find plants just about _____
   on Earth.

   _____
   _ _ _ _ _ _ _ _ _ _

2. Plants _____ where they
   get what they need to live.

   _____
   _ _ _ _ _ _ _ _ _ _

3. Some plants live in the _____
   where it is hot and dry.

   _____
   _ _ _ _ _ _ _ _ _ _

4. These plants can store _____
   in their parts.

   _____
   _ _ _ _ _ _ _ _ _ _

5. Large pointy _____ keep
   rain forest plants from getting too much
   water.

How can plants survive in the cold?

_____

– – – – – – – – – – –

6. Some plants live in the _____

where it is cold and icy.

7. Plants grow close together on the

_____

– – – – – – – – –

_____ to protect them

from the cold and wind.

_____

– – – – – – – – – –

8. Arctic plants also have _____

that grow close to the ground.

Critical Thinking

9. How do plants survive in different places?

_____

– – – – – – – – – – – – – – – – – –

_____

_____

– – – – – – – – – – – – – – – – – –

_____

_____

– – – – – – – – – – – – – – – – – –

_____

Name _____

Plants Live in Many Places

Look at each picture. Circle the word that tells where each plant lives.

1.

desert   rainforest

2.

Arctic   rainforest

3.

Arctic   desert

Plants Live in Many Places

Fill in the blanks. Use the words from the box.

Arctic	desert	parts

Plants can live just about anywhere on

_____

_ _ _ _ _ _ _ _ _ _ _

Earth. They have _____

that help them live in different places.

The roots of plants that live in the

_____

_ _ _ _ _ _ _ _ _ _ _

_____ help them

survive in frozen soil. Plants that live in the

_____

_ _ _ _ _ _ _ _ _ _

_____ have thick stems

that help them hold water. Different plant

parts help plants get what they need to live.

Your Own Garden

✏️ Write About It

Write about a garden that you would like to have. What plants grow in your garden? What do the plants need to grow? Are there plants that you can eat in your garden? Use words that tell what your garden looks like.

Getting Ideas

Picture your garden. It can be a real garden. It can be one you make up. Use the chart below. Write Garden in the center circle. In the outer circles, write the plants that grow in your garden. You can also write plants that you would like to grow.

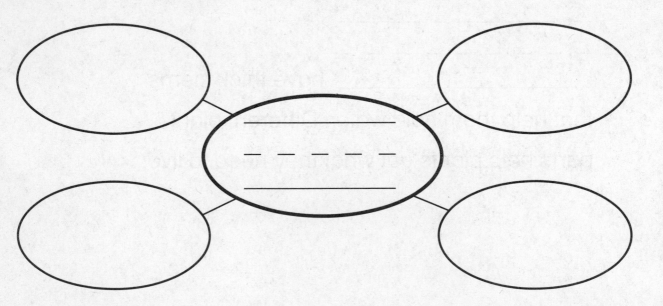

Drafting

Write a sentence. Tell a big idea. Why is your garden special?

_____

— — — — — — — — — — — — — — — —

_____

Now write your story. Begin with the sentence you wrote above. Tell what the garden looks like. Tell what the plants need to grow. Tell about plants you can eat.

_____

— — — — — — — — — — — — — — — —

_____

— — — — — — — — — — — — — — — —

_____

— — — — — — — — — — — — — — — —

_____

Now revise and proofread your writing.
Ask yourself:

▶ Did I name the plants in my garden?

▶ Did I tell what the garden looks like?

▶ Did I correct all mistakes?

Name _____

Plants Grow and Change

Draw a line from the picture to the word that tells about it.

1.

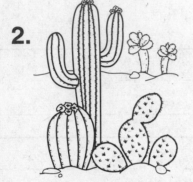

desert

2.

fruit

seed

3.

flower

4.

Unscramble each word. Write it on the line. Use the words in the box below.

| flower | fruit | life cycle | rainforest | seedling |

1. **traosnrife** A _____ is a very rainy place.

2. **gendlsei** A _____ is a very small plant.

3. **rlefwo** A _____ is a plant part that makes seeds.

4. **ifel yccel** A plant's _____ tells how a plant begins, grows, and dies.

5. **urtfi** A _____ is a plant part that protects seeds.

UNIT Literature

Giraffes
by Mary Ann Hoberman

Read the Unit Literature pages in your book.

✏ Write About It
Response to Literature

1. How does the poet tell the reader what giraffes look like?

_____

_ _ _ _ _ _ _ _ _ _ _ _ _ _ _

_____

_ _ _ _ _ _ _ _ _ _ _ _ _ _ _

_____

2. The poet's favorite animal is a giraffe. What is your favorite animal? Write a sentence that tells about it.

_____

_ _ _ _ _ _ _ _ _ _ _ _ _ _ _

_____

_ _ _ _ _ _ _ _ _ _ _ _ _ _ _

_____

All About Animals

Fill in the important ideas as you read the chapter. Use the words in the box.

air	fish	insect	reptile	teeth
bird	food	lungs	scales	water
feathers	gills	mammal	shelter	

All About Animals		
Kinds of Animals	Animal Needs	Animal Body Parts
_____	_____	_____
_____	_____	_____
_____	_____	_____
_____	_____	_____
_____	_____	_____
_____	_____	_____
_____	_____	_____

Name _____

All Kinds of Animals

Use your book to help you fill in the blanks.

What are some types of animals?

1. There are many groups of _____ .

2. Mammals have hair or _____ .

3. The only animals that have feathers are _____ _____ .

What are reptiles and amphibians?

4. Reptiles have dry skin and _____ .

5. Animals that are _____ usually hatch from eggs in the water.

6. Amphibians have _____ , damp skin.

What are some other types of animals?

7. Fish have _____ and

   most have scales.

8. Fish also have _____ to

   help them breathe.

9. Animals that have three body parts and six

   legs are _____ .

Critical Thinking

10. What is one way that birds, reptiles,

    amphibians, fish, and insects are all alike?

    _____

    _____

    _____

    _____

    _____

    _____

Name _____

All Kinds of Animals

Draw a line from the picture to the word that names each animal's group.

1.    insect

2.    fish

3.    bird

4.    mammal

5.    amphibian

6.    reptile

All Kinds of Animals

Fill in the blanks. Use the words from the box.

bird	insect	mammals

Scientists put most animals into

six groups. Animals that have three

body parts and six legs are in the

_____

_ _ _ _ _ _ _

_____ group.

Animals covered with hair or fur are

_____

_ _ _ _ _ _ _

_____ . If you see

an animal with feathers, it is a

_____

_ _ _ _ _ _ _

_____ .

Both fish and reptiles have scales.

Most amphibians have smooth, damp skin.

What Animals Need to Live

Use your book to help you fill in the blanks.

What do animals need to live?

_____

— — — — — — — — — — —

1. Animals are _____ things.

_____

— — — — — — — — — — —

2. Animals need food, _____ ,
   and air.

_____

— — — — — — — — — — —

3. They also need a _____
   place to live.

_____

— — — — — — — — — — —

4. A _____ is a place where
   animals can be safe.

_____

— — — — — — — — — — —

5. Some animals live in _____ ,
   and others live on land.

© Macmillan/McGraw-Hill

How do animals meet their needs?

6. Animals may use their eyes or

_____

– – – – – – – – – –

_____ to find food.

_____

– – – – – – – – – –

7. Fish use _____ to

breathe.

_____

– – – – – – – – –

8. Mammals use _____

to breathe.

Critical Thinking

9. How could you help a dog meet its needs?

_____

– – – – – – – – – – – – – – – – – – – –

_____

– – – – – – – – – – – – – – – – – – – –

_____

– – – – – – – – – – – – – – – – – – – –

_____

Name _____

What Animals Need to Live

Unscramble the word. Then write it on the line.

_____
_ _ _ _ _ _ _ _ _

1. ehletrs A cave can be a _____

for a mountain lion.

_____
_ _ _ _ _ _ _ _ _

2. lilgs A clown fish uses _____

to help take in air from the water.

_____
_ _ _ _ _ _ _ _ _

3. ugnsl A dog uses _____ to help

it breathe the air.

_____
_ _ _ _ _ _ _ _ _

4. inwgs A bird uses its _____

to fly to find food.

What Animals Need to Live

Fill in the blanks. Use the words from the box.

eyes	shelter	wings

All animals need air, food,
and water to live. They also need a

_____

_ _ _ _ _ _ _ _ _

_____ where they
can be safe. Animals use their legs,

_____

_ _ _ _ _ _ _ _ _

fins, or _____ to
move. As they move, they can use their

_____

_ _ _ _ _ _ _ _ _

_____ and noses to find
food and water.

By using their body parts, animals get
what they need to live.

Name _____

Animals' Needs

✏️ **Write About It**

If you have a pet, write about how you care for your pet. Tell what you do first, next, and last. If you don't have a pet, write about a pet you wish you had. Draw a picture to go along with your writing.

Getting Ready

Write the name of your pet in the center of the star. Write what it needs first, next, and last.

_____
_ _ _ _ _ _ _ _ _ _ _ _

first: _____
_____
_ _ _ _ _ _ _ _ _ _ _ _

next: _____
_____
_ _ _ _ _ _ _ _ _ _ _ _

last: _____

© Macmillan/McGraw-Hill

Drafting

The main idea is the most important idea.
Write a main idea about you and your pet.

_____

– –

_____

– –

_____

Now write about you and your pet. Begin with
your main idea sentence. Tell about your pet's
needs.

_____

– –

_____

_____

– –

_____

Revising and Proofreading

Now look at your paragraph. Ask yourself:

▶ Did I begin with my main idea?

▶ Did I include information about the pet's needs?

▶ Did I correct all mistakes?

Name _____

How Animals Eat Food

Use your book to help you fill in the blanks.

Which animals eat plants?

_____

_ _ _ _ _ _ _ _ _ _

1. Animals get _____ from

   the food they eat.

2. An animal that only eats plants is an

   _____

   _ _ _ _ _ _ _ _ _ _

   _____ .

   _____

   _ _ _ _ _ _ _ _ _ _

3. Herbivores have _____

   teeth.

4. An herbivore's teeth help it to chew and

   _____

   _ _ _ _ _ _ _ _ _ _

   _____ plants.

Which animals eat meat?

5. An animal that only eats other animals

   _____

   _ _ _ _ _ _ _ _ _

   is a _____ .

   _____

   _ _ _ _ _ _ _ _ _

6. Carnivores have _____

   teeth.

7. A carnivore's teeth help it to rip and tear

   _____

   _ _ _ _ _ _ _ _ _

   _____ .

Critical Thinking

8. How can scientists figure out what a

   dinosaur ate?

   _____

   _ _ _ _ _ _ _ _ _ _ _ _ _ _

   _____

   _ _ _ _ _ _ _ _ _ _ _ _ _ _

   _____

How Animals Eat Food

Read the sentences. Write YES if the sentence is true. Write NO if the sentence is not true.

_____

1. A carnivore eats meat.

_____

2. A carnivore has flat teeth.

_____

3. A lion is a carnivore.

_____

4. A herbivore eats plants and meat.

_____

5. A horse is a herbivore.

_____

6. A herbivore grinds food with flat teeth.

_____

7. A carnivore has sharp, pointed teeth.

How Animals Eat Food

Fill in the blanks. Use the words from the box.

carnivores	flat	herbivores	sharp

Animals use their teeth to eat food.

_____

_ _ _ _ _ _ _ _ _ _

Animals called _____

eat only meat. These animals have

_____

_ _ _ _ _ _ _ _ _ _

_____ , pointed teeth.

Other animals called

_____

_ _ _ _ _ _ _ _ _ _

_____ eat only

plants. These animals have large,

_____

_ _ _ _ _ _ _ _ _ _

_____ teeth. Some

animals have both kinds of teeth to help

them eat meat and plants.

Name _____

Animals Grow and Change

Use your book to help you fill in the blanks.

How do mammals grow and change?

_____

_ _ _ _ _ _ _ _ _

1. A _____ is all the parts of

an animal's life.

_____

_ _ _ _ _ _ _ _ _

2. Mammals give birth to _____

young.

3. A mammal needs its mother after it is

_____

_ _ _ _ _ _ _ _ _

_____ .

How do birds grow and change?

_____

_ _ _ _ _ _ _ _ _

4. Birds _____ from eggs.

_____

_ _ _ _ _ _ _ _ _

5. Birds start life without _____ .

How do frogs grow and change?

6. Frogs lay eggs in _____ .

7. Young frogs are called _____ .

8. Tadpoles lose their _____

and tails as they grow older.

9. They grow _____ and

legs to become frogs.

Critical Thinking

10. In what ways are all animal life cycles alike?

In what ways are they different?

_____

_____

_____

_____

Name _____

Animals Grow and Change

Complete this diagram. Use the words in the box.

adult frog	eggs	life cycle	tadpole

_____

- - - - - - - - - - - - - - - -

What is a frog's _____ **?**

_____

- - - - - - - - - - - - - - - -

_____

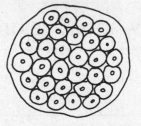

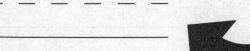

© Macmillan/McGraw-Hill

Animals Grow and Change

Fill in the blanks. Use the words from the box.

fly	hatch	lungs	tadpoles

Animals grow in different ways. Young

_____

– – – – – – – – – –

frogs called _____ hatch

from eggs in water. Later, they lose their gills

_____

– – – – – – – – – – –

and tails and grow _____

and legs.

Birds also lay eggs that

_____

– – – – – – – – – –

_____ into baby birds. As

baby birds grow, they learn to find food and

_____

– – – – – – – – –

_____ . Mammals are

cared for by their mothers until they are older.

Meet Melanie Stiassny

Read the Reading in Science pages in your book. Look for information you can compare and contrast as you read. Remember, when you compare, you tell how things are alike. When you contrast, you tell how things are different. Fill in the diagram below. Tell how an eel is alike and different from a bird.

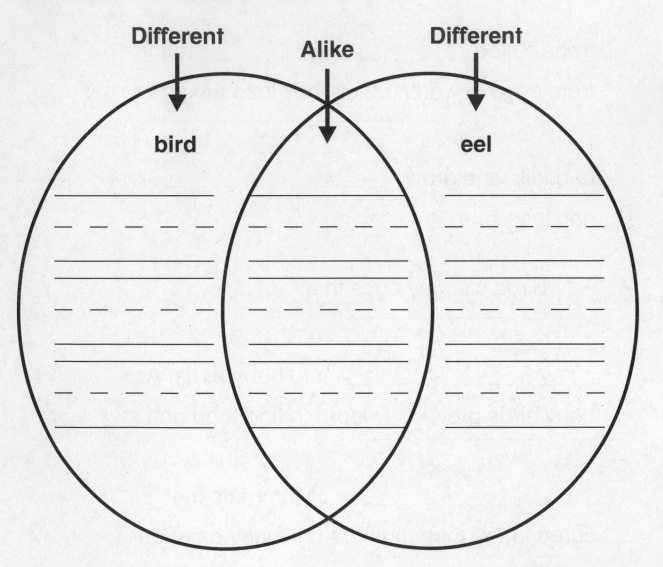

Different   Alike   Different

bird   eel

© Macmillan/McGraw-Hill

Write About It

Compare and Contrast. How can you compare and contrast an eel to another animal you know about? Finish this explanation. Use the words from the box.

fresh	land	river	salty

Eels and frogs are alike and different in some ways. Both life cycles begin in the water. Eels lay their eggs in

_____

_____

_____ water. But frogs lay

_____

_____

their eggs in _____ water.

When eels are adults, they swim into a

_____

_____

freshwater _____ to

live. When frogs are adults, they move onto

_____

_____

_____ .

Name _____

All About Animals

Circle the words that can tell about parts of each picture.

1.

carnivore

herbivore

reptiles

2.

mammals

gills

herbivores

3.

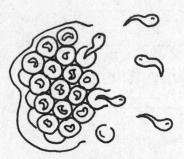

tadpoles

amphibians

insects

4.

mammals

life cycle

bird

Solve the crossword puzzle. Use the words in the box.

| birds | fish | hatch | insects |
| carnivore | gills | herbivore | shelter |

Across

1. I help fish breathe.

3. Chicks do this.

5. I only eat meat.

7. I have feathers.

Down

2. I have six legs.

3. I eat only plants.

4. I'm a safe place to live.

6. I can breathe under water
my whole life.

Name _____

Places to Live

Use the words in the box to tell about a food chain.

animals	food chain	plants	Sun

A _____ includes

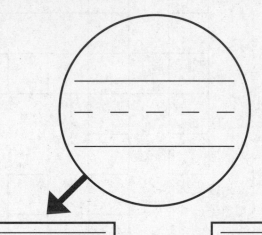

What might happen if there were no Sun?

_____

_____

_____

_____

Land Habitats

Use your book to help you fill in the blanks.

What is a grassland habitat?

_____

— — — — — — — — — —

1. A _____ is the place

   where an animal lives.

   _____

   — — — — — — — — —

2. A _____ is a land habitat

   that is dry with a lot of grass.

3. Animals have adaptations that help them

   _____

   — — — — — — — — —

   _____ in their habitat.

   _____

   — — — — — — — —

4. An _____ is a body feature

   or behavior that helps an animal stay safe.

   _____

   — — — — — — — —

5. A giraffe's long _____

   is an adaptation that helps it stay safe.

What is a forest habitat?

_____

– – – – – – – – – – – –

6. A tree can be a home for both _____ and plants.

_____

– – – – – – – – – – – –

7. There are many trees in a _____ .

8. Some trees grow tall to help them get _____

– – – – – – – – – – –

_____ .

_____

– – – – – – – – – – –

9. Some animals use trees for _____ or eat nuts or insects found on trees.

Critical Thinking

10. Some rabbits turn white in winter. How would this adaptation help them?

_____

– – – – – – – – – – – – – – – – – – –

_____

– – – – – – – – – – – – – – – – – – –

Land Habitats

Look at the pictures. Write about where each animal lives and how it gets what it needs. Use each word in the box at least once.

adaptation	forest	grassland	habitat

1.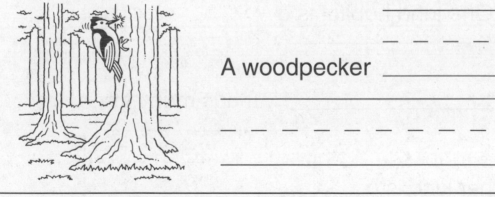

A giraffe's neck _____

_____

_____

2.

A woodpecker _____

_____

_____

3.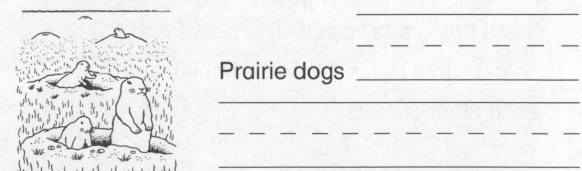

Prairie dogs _____

_____

_____

Land Habitats

Fill in the blanks. Use the words from the box.

forest	grassland	habitat

Different plants and animals live in
different places. Each plant and animal has

_____

_ _ _ _ _ _ _ _ _

its own _____ where it can

meet its needs.

One land habitat is a

_____

_ _ _ _ _ _ _ _ _

_____ . It has many trees.

_____

_ _ _ _ _ _ _ _ _ _

Another kind is a _____ .

It is a dry, grassy place. The plants and

animals that live in the same habitat help

each other survive.

Water Habitats

Use your book to help you fill in the blanks.

What is a lake habitat?

_____
_ _ _ _ _ _ _ _ _

1. A _____ is a body of
fresh water with land on all sides.

_____
_ _ _ _ _ _ _ _ _

2. Fresh water has little or no _____ .

_____
_ _ _ _ _ _ _ _

3. A lake is a _____ habitat
where plants and animals live.

_____
_ _ _ _ _ _ _ _ _

4. Many animals find food and _____
in lakes.

What is an ocean habitat?

5. Another kind of water habitat is an
_____
_ _ _ _ _ _ _ _
_____ .

Name _____

6. An ocean is a large, deep body of

_____

– – – – – – – – – –

_____ water.

_____

– – – – – – – – –

7. Many different _____ ,

fish, and plants live in the ocean and help

each other survive.

8. Some animals, like whales, eat small

_____

– – – – – – – – – –

_____ , and other animals

eat plants in the ocean.

_____

– – – – – – – –

9. Many animals find _____

and shelter in oceans.

Critical Thinking

10. Could an ocean fish live in a lake? Why?

_____

– –

_____

_____

– –

_____

Water Habitats

Look at the pictures. Circle the correct word for each. Then write to describe each water habitat.

I.

lake   ocean

_____
- - - - - - - - - - - - - - - - - - -
_____
- - - - - - - - - - - - - - - - - - -
_____
- - - - - - - - - - - - - - - - - - -
_____

2.

lake   ocean

_____
- - - - - - - - - - - - - - - - - - -
_____
- - - - - - - - - - - - - - - - - - -
_____
- - - - - - - - - - - - - - - - - - -
_____

Water Habitats

Fill in the blanks. Use the words from the box.

fresh	lake	ocean	salty

Living things are found in different water

_____

_ _ _ _ _ _ _ _ _

habitats. The _____ is

the largest water habitat. Its

_____

_ _ _ _ _ _ _ _ _

_____ water is home to

many plants and animals.

_____

_ _ _ _ _ _ _ _ _

A _____ is

much smaller than the ocean. Its

_____

_ _ _ _ _ _ _ _ _

_____ water has little or

no salt. The plants and animals who live there

depend on each other to survive.

Arctic Adaptions

✏️ Write About It

Write about how the arctic fox survives in the arctic.

Who?

Who is the animal?

_____
_ _ _ _ _ _ _ _ _ _ _ _ _
_____

What?

What does it eat?

_____
_ _ _ _ _ _ _ _ _ _ _ _ _
_____

When?

When does it do this?

_____
_ _ _ _ _ _ _ _ _ _ _ _ _
_____

Where?

Where does it do this?

_____
_ _ _ _ _ _ _ _ _ _ _ _ _
_____

How?

How does it do it?

_____
_ _ _ _ _ _ _ _ _ _ _ _ _
_____

Drafting

The main idea is the most important idea.
Write a main idea about the arctic fox.

_____

_ _

_____

_ _

_____

Now write your paragraph.

_ _

_____

_ _

_____

_____

_ _

Revising and Proofreading

Now look at your paragraph. Ask yourself:

▶ Did I begin with a main idea?

▶ Did I describe what the arctic fox eats?

▶ Did I correct all spelling, punctuation, and capital
letter mistakes?

Plants and Animals Live Together

Use your book to help you fill in the blanks.

Why do plants and animals live together?
_____
_ _ _ _ _ _ _ _ _ _

1. Plants help _____ live.

_____
_ _ _ _ _ _ _ _ _

2. Animals use plants for _____
and food.

_____
_ _ _ _ _ _ _ _ _

3. Bees help plants make _____
plants by carrying pollen.

What is a food chain?
_____
_ _ _ _ _ _ _ _ _

4. A _____ shows the order
in which living things get food.

_____
_ _ _ _ _ _ _ _ _

5. The _____ is at the
beginning of every food chain.

_ _ _ _ _ _ _ _ _

6. Plants are the _____ living

link in most food chains.

What happens to living things when a habitat changes?

7. A habitat can be changed by people, animals,

_ _ _ _ _ _ _ _ _ _ _

plants and _____ .

_ _ _ _ _ _ _ _ _

8. A living thing becomes _____

when there are no more of its kind on Earth.

Critical Thinking

9. Where are animals found in most food chains?

_ _ _ _ _ _ _ _ _ _ _ _ _ _ _ _

_____

_____

_ _ _ _ _ _ _ _ _ _ _ _ _ _ _ _

_____

Plants and Animals Live Together

Read the sentences about plants and animals. Write YES if the sentence is true. Write NO if the sentence is not true.

_____

1. All living things need food to give them energy.

_____

2. A food chain shows the order in which living things get the food they need.

_____

3. The Sun is the first link in food chains.

_____

4. People are at the top of many food chains.

_____

5. A change in a habitat can bring back a living thing that is extinct.

Plants and Animals Live Together

Fill in the blanks. Use the words from the box.

food chain	plants
people	Sun

All animals depend on plants in order

_____

_ _ _ _ _ _ _ _ _ _

to live. A _____ shows

how energy passes from plants to animals.

_____

_ _ _ _ _ _ _ _ _

The _____ is at the

beginning of all food chains. The Sun gives

_____

_ _ _ _ _ _ _ _ _

_____ energy to make food.

_____

_ _ _ _ _ _ _ _ _

When _____ eat plants

or foods from animals, this energy is passed

on. Without the Sun, all living things would

become extinct.

Meet Jin Meng

Read the Reading in Science pages in your book. Look for information that shows cause and effect as you read. Remember, cause is what makes something happen. Effect is something that is produced by a cause. Fill in the diagram below. Tell how a dinosaur's teeth affects what it eats.

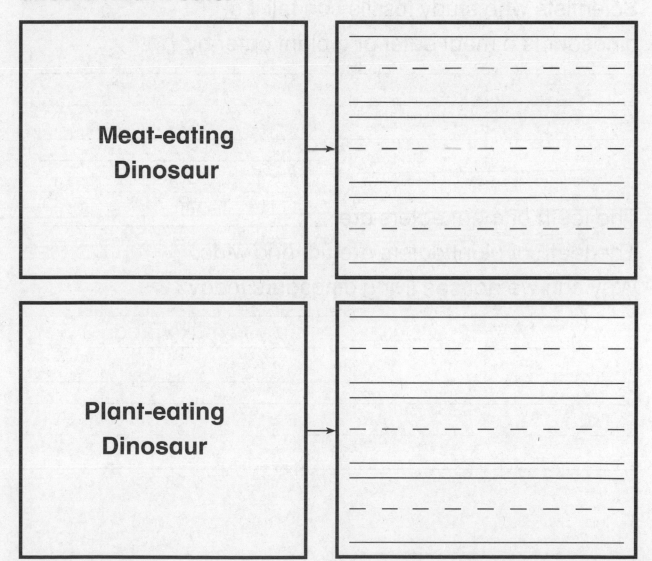

Meat-eating Dinosaur

Plant-eating Dinosaur

✏ Write About It

Cause and Effect. How does a scientist who studies fossils recognize a meat-eating dinosaur? Finish this explanation. Use the Cause and Effect diagram you made on page 79.

Scientists who study fossils can tell if a dinosaur is a meat eater or a plant eater by

_____

_ _ _ _ _ _ _ _ _ _ _ _ _ _ _ _

_____ .

_____

_ _ _ _ _ _ _ _ _ _

The teeth of meat eaters are _____ .
The teeth of plant eaters are flat and wide.
Why can we not see living dinosaurs today?

_____

_ _ _ _ _ _ _ _ _ _ _ _ _ _ _ _

_____

_____

_ _ _ _ _ _ _ _ _ _ _ _ _ _ _ _

_____

Places to Live

Solve the crossword puzzle. Use the chapter vocabulary words from your book.

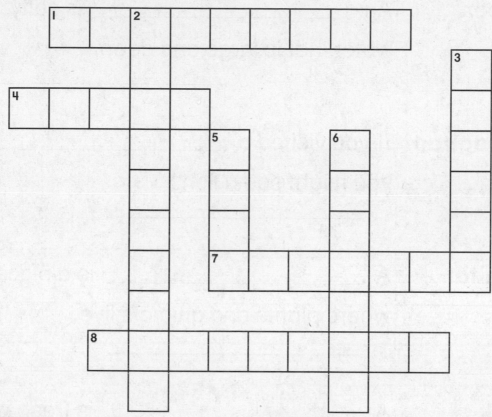

Across

1. Dry place with grass

4. Large, deep body of salt water

7. Die out forever

8. Shows what animals eat

Down

2. Body part that helps a living thing survive

3. Place with lots of trees

5. Body of fresh water

6. Place where an animal lives

Unscramble each word. Write it on the line to finish the sentence.

1. **caoen**

   _____

   _ _ _ _ _ _ _ _

   An _____ is salty water that is large and deep.

2. **dssaglran**

   _____

   _ _ _ _ _ _ _ _ _

   If you visited a _____ , you might see a lion.

3. **abhitat**

   _____

   _ _ _ _ _ _ _ _

   A _____ is a place where plants and animals live.

4. **aelk**

   _____

   _ _ _ _ _ _ _ _

   A _____ is fresh water with land around it.

5. **xncetti**

   Dinosaurs are an example of an

   _____

   _ _ _ _ _ _ _

   _____ animal.

Sand

by Meish Goldish

Read the Unit Literature pages in your book.

✏ Write About It

Response to Literature

1. How do you think the author feels about
   sand? Why?

   _____

   _

   _____

   _

   _____

   _

   _____

2. The author tells about many places where
   there is sand. On a separate sheet of paper,
   draw a picture of one of the places with sand
   the author talks about.

Looking at Earth

Fill in the important ideas as you read the chapter. Use the words in the box.

continents	mountains	soil
erosion	ocean	weathering
lakes	rivers	

What do water and land look like?

Earth's Land

_____

_ _ _ _ _ _ _ _ _ _ _ _ _

_____

_ _ _ _ _ _ _ _ _ _ _ _ _

_____

Earth's Water

_____

_ _ _ _ _ _ _ _ _ _ _ _ _

_____

_ _ _ _ _ _ _ _ _ _ _ _ _

_____

Earth Changes

_____   _____

_ _ _ _ _ _ _ _ _ _ _ _ _   _ _ _ _ _ _ _ _ _ _ _ _ _

_____   _____

© Macmillan/McGraw-Hill

What Earth Looks Like

Use your textbook to help you fill in the blanks.

What is on the surface of Earth?

_____
– – – – – – – – – – – –

1. Earth is made of _____
and mostly water.

_____
– – – – – – – – – – –

2. Earth has _____ big
pieces of land called continents.

_____
– – – – – – – – – – – –

3. Water surrounds the _____ .

What is Earth's water like?

_____
– – – – – – – – – –

4. Most of _____'s water is
in oceans.

_____
– – – – – – – – – –

5. An _____ is a big and
deep body of salt water.

_ _ _ _ _ _ _ _

6. Earth's fresh water is in streams, _____ ,
_____

_ _ _ _ _ _ _ _

and _____ .

What is Earth's land like?

_____
_ _ _ _ _ _ _ _

7. Some of Earth's land, like _____ ,
is very high.

_____
_ _ _ _ _ _ _ _

8. Some land is low or flat, like a _____
or plains.

Critical Thinking

9. How would you describe Earth to a new friend?

_ _

_____

_____

_ _

_____

_____

_ _

_____

What Earth Looks Like

Circle the words that tell about Earth in the puzzle below.

| continent | mountains | rivers |
| lakes | plains | stream |

```
c  o  n  t  i  n  e  n  t
w  o  a  o  t  i  z  k  n
a  u  i  p  p  b  w  h  o
s  e  w  l  a  k  e  s  g
t  h  x  a  u  k  o  j  j
r  c  r  i  v  e  r  s  o
e  b  l  n  e  x  c  q  p
a  w  p  s  i  u  d  z  v
m  o  u  n  t  a  i  n  s
```

Name _____

What Earth Looks Like

Fill in the blanks. Use the words from the box.

lakes	ocean	rivers

Much of Earth's water is in oceans. Most

people and animals can not drink salty water

_____

_ _ _ _ _ _ _ _ _

from the _____ .

_____

_ _ _ _ _ _ _ _ _

Earth has _____ ,

lakes and streams, too. Some rivers flow into

_____

_ _ _ _ _ _ _ _ _

_____ or the ocean. A

valley is a low place between mountains.

Rocks and Soil

Use your textbook to help you fill in the blanks.

What are rocks?

1. Rocks can look and feel _____ .

2. Some rocks are _____ and shiny.

3. Others rocks are _____ and dull.

4. Some _____ are made of only one mineral.

5. Other rocks are made of _____ minerals.

6. A mineral _____

What is soil?

_____
_ _ _ _ _ _ _ _

7. The top layer of _____ is
called soil.

8. It is made of tiny pieces of rocks and dead
_____
_ _ _ _ _ _ _ _
_____ and animals.

_____
_ _ _ _ _ _ _ _

9. Air and water are also in _____ .

10. Soils can look different when there are
_____
_ _ _ _ _ _ _ _
_____ plants, animals and
rocks in them.

Critical Thinking

11. Compare rocks and soil. How are they alike?
How are they different?

_____
_ _ _ _ _ _ _ _ _ _ _ _
_____
_____
_ _ _ _ _ _ _ _ _ _ _ _
_____

Rocks and Soil

Unscramble each word. Use it to complete the sentences.

limensra

1. Rocks are different when they are made of

_____

_ _ _ _ _ _ _ _

different _____ .

_____

_ _ _ _ _ _ _ _

2. Some _____ make

rocks hard.

lsoi

3. Tiny bits of rock, dead plants, and animals

_____

_ _ _ _ _ _ _ _ _

make up _____ .

4. Most plants do not grow well in clay

_____

_ _ _ _ _ _ _ _

_____ .

Rocks and Soil

Fill in the blanks. Use the words from the box.

Earth	layer	plants	rocks

Rocks and soil have a lot in common. They are both very important to

_____

_ _ _ _ _ _ _ _ _

_____ . All rocks are

made of minerals. The top

_____

_ _ _ _ _ _ _ _

_____ of Earth is called soil.

_____

_ _ _ _ _ _ _ _ _ _

Trees and _____ grow

in soil. Many animals also live in soil.

Soil is made up of tiny bits of

_____

_ _ _ _ _ _ _ _

_____ . It is also made

of tiny bits of dead plants and animals, air,

and water.

Meet Rondi Davies

Read the Reading in Science pages in your book. Look for the order in which things happen as you read. Fill in the diagram below. Tell what happens first, next, and last when diamonds are formed.

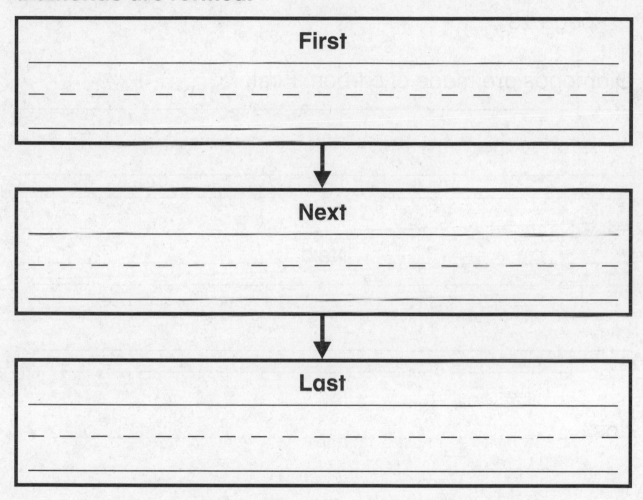

First

_____

- -

_____

Next

_____

- -

_____

Last

_____

- -

_____

Write About It

Put Things in Order. How do diamonds form? Put things in order to finish this explanation. Use the chart you made that tells the order in which things happen on page 93.

Diamonds are made of carbon. First,

_____

- - - - - - - - - - - - - - - -

_____

_____   _____

- - - - - - - -   - - - - - - - -

_____ . Next, _____

_____

- - - - - - - - - - - - - - - -

_____ .

_____

- - - - - - - - - - - - - - - -

Last, _____ .

Changing the Land

Use your textbook to help you fill in the blanks.

How can rocks change?

_____

– – – – – – – – – –

1. Rocks can _____ size
   and shape.

_____

– – – – – – – – – –

2. Water can _____ and
   break apart rocks over time.

_____

– – – – – – – – – –

3. This is called _____ .

_____

– – – – – – – – – –

4. Plants can also _____
   rocks through weathering.

_____

– – – – – – – – – –

5. Sometimes _____ will
   grow into the cracks of rocks.

How can land change?

6. Land changes shape slowly through

_____

_ _ _ _ _ _ _ _ _ _

_____ .

_____

_ _ _ _ _ _ _ _ _ _ _

7. Wind and _____ help to

change soil or rocks over time through erosion.

_____

_ _ _ _ _ _ _ _ _

8. A _____ is formed when

water slowly washes away pieces of rock.

9. Plants help stop erosion by holding

_____

_ _ _ _ _ _ _ _ _ _

_____ in place.

Critical Thinking

10. How can Earth's land change over time?

_____

_ _ _ _ _ _ _ _ _ _ _

_____

_ _ _ _ _ _ _ _ _ _

_____

Changing the Land

Look at the pictures, then read the sentence next to each one. Write TRUE if the sentence is true. Write FALSE if the sentence is not true.

I.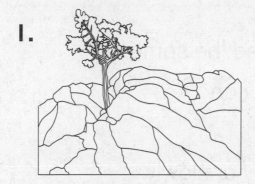

The roots of this tree are pushing against the rock.

_____

– – – – – – –

_____

2.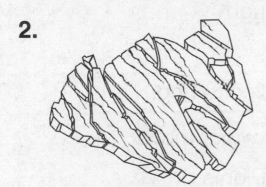

Water is changing the shape of this rock through erosion.

_____

– – – – – – –

_____

3.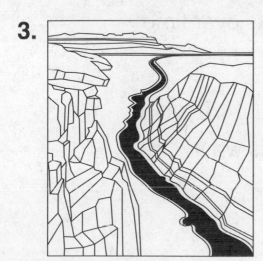

When wind and water move pieces of rocks and soil to a new place it is called erosion.

_____

– – – – – – –

_____

Changing the Land

Fill in the blanks. Use the words from the box.

change	erosion	weathering

Earth has not always looked the same.
It has changed over time. Earth can
_____

— — — — — — — — —

_____ quickly or slowly.
Earth changes very slowly through
_____

— — — — — — — — —

_____ and weathering.

Erosion is when wind and water move
rocks and soil away over time. Plants or
water change the shape of rocks over time
_____

— — — — — — — — —

through _____ . It takes
many years to change Earth this way!

Stopping Erosion

✏ Write About It

Look at this picture. What could be eroding the soil here? Write a story about how erosion could be stopped in this picture.

Getting Ideas

Use the web below. Write Erosion in the center circle. Fill in the other circles. Write things you see to stop erosion.

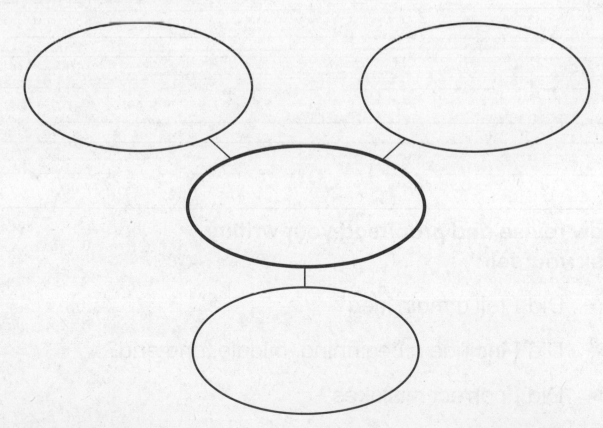

© Macmillan/McGraw-Hill

Drafting

The main idea is the most important idea.
Write the main idea for your story.

_____

- -

_____

_____

- -

_____

Now write your story. Begin with the sentence
you just wrote. Tell how erosion could be
stopped in the picture. Include a beginning,
middle, and end.

_____

- -

_____

_____

- -

_____

_____

- -

_____

Now revise and proofread your writing.
Ask yourself:

► Did I tell a main idea?

► Did I include a beginning, middle, and end?

► Did I correct mistakes?

Our Earth

Write the word that best describes each statement.

continents	mountain	soil

1.

This picture shows the biggest pieces of land on Earth.

_____

- - - - - - - - - - - -

_____

2.

This picture shows something that is made of minerals and dead plants and animals.

_____

- - - - - - - - - - - -

_____

3.

This picture shows land that is high above the ocean.

_____

- - - - - - - - - - - -

_____

Find the words in the box to answer the statements. Write the correct word on the line.

lakes	rivers	weathering
ocean	valley	

4. This is big and salty. It takes up most of Earth.

_____

_ _ _ _ _ _ _

_____

5. This is how plants and water change the

shape of rocks.

_____

_ _ _ _ _ _ _ _

_____

6. Two fresh water places.

_____   _____

_ _ _ _ _ _   _ _ _ _ _ _

_____ , _____

7. Between two mountains.

_____

_ _ _ _ _ _ _

_____

Caring for Earth

Fill in the important ideas as you read the chapter. Use the words in the box.

air	people	recycle	reuse
land	pollution	reduce	water

Earth's Resources

Types of Natural Resources a Seed Needs to Grow

What Harms Earth?

Ways to Conserve Resources

_____
- - - - - - - - - - - - -
_____
- - - - - - - - - - - - -
_____

_____
- - - - - - - - - - - - -
_____
- - - - - - - - - - - - -
_____

_____
- - - - - - - - - - - - -
_____
- - - - - - - - - - - - -
_____

Earth's Resources

Use your book to help you fill in the blanks.

What is a natural resource?

_____

_ _ _ _ _ _ _ _ _ _

1. A _____ is something

   people use that comes from Earth.

   _____

   _ _ _ _ _ _ _ _ _ _ _

2. Rocks, soil, _____ , and

   air are kinds of natural resources.

   _____

   _ _ _ _ _ _ _ _ _

3. Plants and _____ are

   natural resources, too.

4. People can use different natural resources in

   _____

   _ _ _ _ _ _ _ _ _

   _____ ways.

Why is soil important?

5. Earth's _____ is an

important natural resource.

6. Plants _____ in soil.

7. People and animals use plants that grow in

soil for _____ .

8. People use soil, like _____ ,

to make things.

Critical Thinking

9. Why is soil an important natural resource?

_____

_____

_____

_____

Name _____

Earth's Resources

Look at the picture. Circle objects that show natural resources. Then write the objects.

_____   _____

_____   _____

_____   _____

_____   _____

Earth's Resources

Fill in the blanks. Use the words from the box.

air	living	natural resource

People use natural resources in many ways. Some of these resources are from

_____

_____ things, like

plants and animals. Others, like water,

_____

_____ , rocks, and soil

are nonliving things.

_____

_____

One important _____ is soil. Soil is used by both people and animals.

Name _____

Using Earth's Resources

Use your book to help you fill in the blanks.

Why are water and air important?

_____
– – – – – – – – –

1. Living things on Earth need _____
   to live.

   _____
   – – – – – – – –

2. Plants need water to _____
   and to make food.

3. Animals and plants use water to
   _____
   – – – – – – – – –
   _____ and clean.

   _____
   – – – – – – – –

4. We need _____ to
   breathe.

   _____
   – – – – – – – –

5. All living things need _____
   air and water to live.

What is pollution?

6. Pollution is when there are

_____

_ _ _ _ _ _ _ _

_____ things in the air,

land, and water.

7. Pollution happens when

_____

_ _ _ _ _ _ _ _

_____ or dirt gets into the

ground, water, or air.

_____

_ _ _ _ _ _ _ _ _

8. Living things can get _____

if soil, water, or air are not clean.

_____

_ _ _ _ _ _ _ _

9. There are _____ ways to

keep Earth clean.

Critical Thinking

10. Do you think it is important to prevent air and

water pollution? Why or why not?

_____

_ _ _ _ _ _ _ _ _ _ _ _ _ _ _ _

_____

Name _____

Using Earth's Resources

Finish each sentence. Tell what kind of pollution is shown in each picture.

_____

I. This picture shows _____ .

_____

2. This picture shows _____ .

© Macmillan/McGraw-Hill

Using Earth's Resources

Fill in the blanks. Use the words from the box.

air	harmful	clean	water

Air and water are important natural resources. Plants must have

_____

_____ and water to grow. Land animals and people need to breathe

_____

_____ air. All animals and people need clean

_____

_____ to drink.

Sometimes air and water become

_____

_____ polluted with _____ things. All living things need clean air and water to survive.

Name _____

Meet Mark Siddall

Read the Reading in Science pages in your book. Think about Problem and Solution as you read the information. Remember, a solution is the answer to a problem. Fill in the chart below.

Where do leeches belong?

Problem	Solution

✏ Write About It

Problem and Solution. What is another animal that can live in water? Draw the animal. What can we do to keep the animal safe? Use what you know and ideas from the Problem and Solution chart you made.

A _____ lives in water.

_____

_____

_____

_____

Saving Earth's Resources

Use your book to help you fill in the blanks.

How can we reuse resources?

1. We protect our natural resources when we
_____

_ _ _ _ _ _ _ _ _ _

_____ them.

2. Another way to save our natural resources is
_____

_ _ _ _ _ _ _ _ _

to _____ them.

3. When you reuse things, you use them again
_____

_ _ _ _ _ _ _ _ _ _

in a _____ way.

_____

_ _ _ _ _ _ _ _ _

4. You do not have to _____

new things when you reuse things.

How can we save resources?

_____

– – – – – – – – – – – –

5. You _____ when you use

less of some things.

_____

– – – – – – – – – – – –

6. When you recycle, you turn _____

things into new things.

_____

– – – – – – – – – –

7. People can _____ paper,

plastic, metal, and glass.

Critical Thinking

8. What can you do to help conserve natural

resources?

_____

– – – – – – – – – – – – – – – – – –

_____

– – – – – – – – – – – – – – – – – –

_____

_____

– – – – – – – – – – – – – – – – – –

_____

Name _____

Saving Earth's Resources

Write the word that fits each meaning. Use the words in the box.

conserve	recycle	reduce	reuse

_____

- - - - - - - - - - -

_____   **1.** to use less of something

- - - - - - - - - - -

_____   **2.** to save, keep or protect
                                     natural resources

_____

- - - - - - - - - - -

_____   **3.** to use something again
                                     in a new way

_____

- - - - - - - - - - -

_____   **4.** to turn something into a
                                     new thing

Name _____

Saving Earth's Resources

Fill in the blanks. Use the words from the box.

conserve	recycle	reduce	reuse

What can you do to show you care

about the Earth? One important thing is to

_____

_ _ _ _ _ _ _ _ _

_____ natural

resources. When you use less, you

_____

_ _ _ _ _ _ _ _ _

_____ the amount of

resources you use. Another thing you can

_ _ _ _ _ _ _ _ _

do is to _____

things again and again. You can

_____

_ _ _ _ _ _ _ _ _

_____ resources like

plastic, paper, glass, and metal. These are

all important ways to show that you care

about Earth's resources.

Name _____

Saving Water

Tell what happens first, next, and last.

✏ Write About It

Write about other ways water might be wasted. Tell what you can do to save water.

Getting Ideas

Choose one way you waste water. Fill in the chart. Tell steps to save water.

```
┌─────────────────────────────────────────┐
│                  First                    │
│  _____  │
│  - - - - - - - - - - - - - - - - - - - -  │
│  _____  │
└─────────────────────────────────────────┘
                      │
                      ▼
┌─────────────────────────────────────────┐
│                  Next                     │
│  _____  │
│  - - - - - - - - - - - - - - - - - - - -  │
│  _____  │
└─────────────────────────────────────────┘
                      │
                      ▼
┌─────────────────────────────────────────┐
│                  Last                     │
│  _____  │
│  - - - - - - - - - - - - - - - - - - - -  │
│  _____  │
└─────────────────────────────────────────┘
```

© Macmillan/McGraw-Hill

Drafting

Begin your story. Write a sentence. Tell what you can do to save water.

_____

_ _ _ _ _ _ _ _ _ _ _ _ _ _ _ _ _ _

_____

_ _ _ _ _ _ _ _ _ _ _ _ _ _ _ _ _ _

Now write your story. Begin with the sentence you wrote. Then tell the steps.

_____

_ _ _ _ _ _ _ _ _ _ _ _ _ _ _ _ _ _

_____

_ _ _ _ _ _ _ _ _ _ _ _ _ _ _ _ _ _

_____

_ _ _ _ _ _ _ _ _ _ _ _ _ _ _ _ _ _

_____

Now look at your paragraph. Ask yourself:

▶ Did I tell how to save water?

▶ Did I put the steps in order?

▶ Did I correct all mistakes?

Name _____

Caring for Earth

Circle the words in each box that tell about the word at the top of the box.

1. pollution smoke flowers litter oil spill	**2. conserve** reuse recycle reduce air
3. natural resource plants animals cars water	**4. recycle** life cycle glass paper plastic

Circle the word that tells how Earth's natural resources are being helped or harmed.

1. reuse   reduce

2. pollution   recycle

3. reuse   pollution

4. recycle   reduce

Name _____

Weather and Animals

Read the Unit Literature pages in your book.

✏ Write About It

Response to Literature

1. What kinds of weather does the article tell about?

_____

- -

_____

- -

_____

2. What do you do in hot and sunny weather?

_____

- -

_____

_____

- -

_____

Weather and Seasons

Read the name of each season. Finish each sentence by telling what you can predict about the weather in each season.

Spring

_____

Spring has _____

_____

- - - - - - - - - - -

_____

- - - - - - - - - - -

_____

Summer

_____

- - - - - - - - - - -

Summer has _____

_____

- - - - - - - - - - -

_____

- - - - - - - - - - -

_____

Winter

_____

- - - - - - - - - - -

Winter has _____

_____

- - - - - - - - - - -

_____

- - - - - - - - - - -

_____

Fall

_____

- - - - - - - - - - -

Fall has _____

_____

- - - - - - - - - - -

_____

- - - - - - - - - - -

_____

Name _____

Weather All Around Us

Use your book to help you fill in the blanks.

What is weather?

1. What the air and sky are like each day is

   _____

   _ _ _ _ _ _ _ _

   _____ .

2. The air might be   _____

   _ _ _ _ _ _ _ _ _ ,

   snowy, or dry.

3. The sky can be   _____

   _ _ _ _ _ _ _ _

   or sunny.

4. The Sun causes the   _____

   _ _ _ _ _ _ _ _

   to change.

5. Wind is   _____   air.

   _ _ _ _ _ _ _ _

How can you measure weather?

6. Temperature can be measured with a

_____

— — — — — — — —

_____ .

_____

— — — — — — —

7. You can use a _____

to measure how much rain falls.

8. You can measure the direction of the wind

_____

— — — — — — — —

with a _____ .

Critical Thinking

9. What things other than a wind vane can help

you figure out the direction of the wind?

_____

— — — — — — — — — — — — — — — —

_____

— — — — — — — — — — — — — — — —

_____

— — — — — — — — — — — — — — — —

_____

Weather All Around Us

Unscramble each word. Write it on the line.

1. ndiw aven To tell wind direction, use a

_____

_ _ _ _ _ _ _

_____ .

2. eumtpteearr If you know how warm or cold

the air is, you know the

_____

_ _ _ _ _ _ _ _ _ _

_____ .

3. anri ugeag To measure rain, use a

_____

_ _ _ _ _ _ _ _

_____ .

4. aewtehr You can tell the

_____

_ _ _ _ _ _ _

_____ by what the

air and sky are like.

5. methermtoer To measure the temperature of the air,

_____

_ _ _ _ _ _ _

use a _____ .

Weather All Around Us

Fill in the blanks. Use the words from the box.

rain gauge	temperature	thermometer	wind vane

You can measure weather in different ways. You can feel the air

_____

_____ to tell if it

is warm or cold. You can also use a

_____

_____ to measure

the temperature. You can use a

_____

_____ to tell the wind's

_____

_____

direction. A _____ will tell

you how much rain has fallen. All these tools

help tell about weather.

Name _____

The Water Cycle

Use your book to help you fill in the blanks.

What makes it rain or snow?

1. The Sun's heat can turn water into
   _____
   _ _ _ _ _ _ _ _
   _____ .

2. Water vapor is water that goes up in the
   _____
   _ _ _ _ _ _ _ _
   _____ .

3. When water vapor _____ ,
   _ _ _ _ _ _ _ _ _
   it turns into drops of water or bits of ice.
   _____
   _ _ _ _ _ _ _ _

4. The _____ are made of
   water and bits of ice.

5. When the water drops or bits of ice get big, they
   _____
   _ _ _ _ _ _ _ _
   can fall as rain, snow, or _____ .

What are some different kinds of clouds?

6. Some clouds look _____ and others look puffy.

7. Cirrus clouds are made of thin bits of _____ _ _ _ _ _ _ _ _ _____ .

8. Cumulus clouds are mostly _____ drops.

9. Cumulonimbus clouds bring _____ and storms.

Critical Thinking

10. What would you wear if you saw cumulonimbus clouds in the sky? Why?

_____

_____

_____

_____

Name _____

The Water Cycle

**Look at the diagram of the water cycle.
Label the parts. Use the words in the box.**

clouds	Sun	water vapor
rain	water	

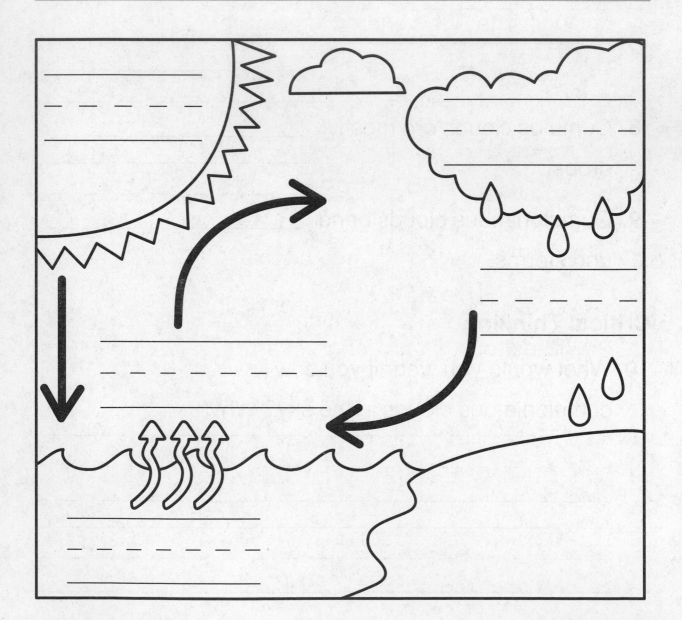

© Macmillan/McGraw-Hill

The Water Cycle

Fill in the blanks. Use the words from the box.

colder	ice	water vapor

The Sun helps make wind, rain, snow,
and hail. The Sun's energy turns water

_____

_ _ _ _ _ _ _ _ _

into _____ that goes
up into the air. When water vapor cools,
it becomes drops of water or tiny bits of

_____

_ _ _ _ _ _ _ _ _

_____ that form clouds.

The drops get bigger or

_____

_ _ _ _ _ _ _ _ _

_____ and fall to Earth as
rain or snow. This is called the water cycle.

Spring and Summer

Use your book to help you fill in the blanks.

What happens in spring?

1. A time of year is a _____ .
   ___ ___ ___ ___ ___ ___ ___ ___

2. The four _____ are
   ___ ___ ___ ___ ___ ___ ___
   winter, spring, summer, and fall.

3. There are more hours of sunlight in
   _____
   ___ ___ ___ ___ ___ ___ ___
   _____ .

4. The rain and extra _____
   ___ ___ ___ ___ ___ ___ ___ ___
   in spring help plants grow.

5. Many _____ are born in
   ___ ___ ___ ___ ___ ___ ___ ___
   spring.

What happens in summer?

_____

_ _ _ _ _ _ _ _ _ _

6. The season after spring is _____ .

_____

_ _ _ _ _ _ _ _ _

7. Summer is the _____

season.

_____

_ _ _ _ _ _ _ _ _ _

8. Many plants grow _____

in summer.

9. During summer, there is more

_____

_ _ _ _ _ _

_____ for animals to eat.

Critical Thinking

10. When do most people go to the beach?

Why?

_____

_ _ _ _ _ _ _ _ _ _ _ _ _ _

_____

_ _ _ _ _ _ _ _ _ _ _ _ _

_____

Spring and Summer

Read the sentences. Write YES if the sentence is true. Write NO if the sentence is not true.

_____
_____

1. A **season** is a time of year.

_____
_____

2. There are three **seasons** of the year.

_____
_____

3. There are more hours of sunlight in **spring** than in winter.

_____
_____

4. In **spring**, plants begin to sprout and many animals are born.

_____
_____

5. **Summer** is the coolest season.

_____
_____

6. Sunlight in **summer** helps plants grow fruits.

© Macmillan/McGraw-Hill

Spring and Summer

Fill in the blanks. Use the words from the box.

daylight	plants	spring

In many places, the weather changes
during the four seasons. The weather begins
_____
_ _ _ _ _ _ _ _
to warm up in _____ .
There are more hours of
_____
_ _ _ _ _ _ _ _
_____ . The rain in spring
_____
_ _ _ _ _ _ _ _
helps _____ grow.

In hot summer weather, people may sit
in the shade or go for a swim.

Museum Mail Call

Read the Reading in Science pages in your book. Fill in the diagram below. Write the important ideas in the small boxes. Then retell these ideas in the big box.

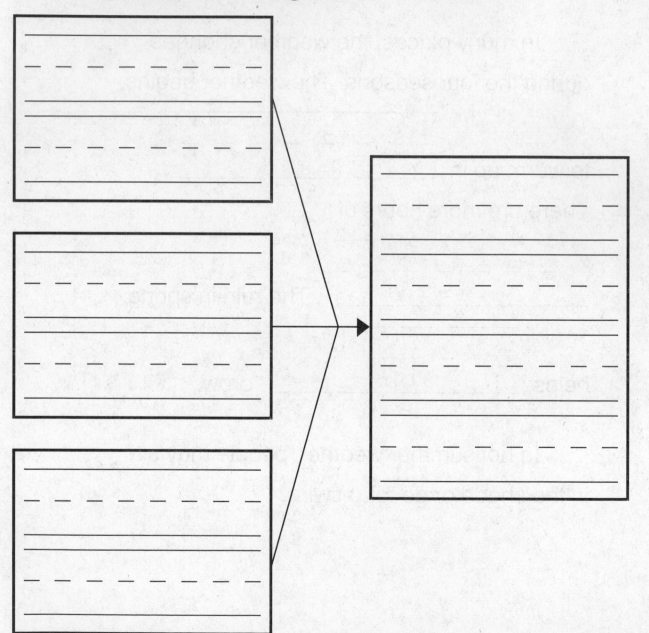

✏️ Write About It

Find Main Idea and Details. What happens in spring to help the Hunza farmers' seeds grow? Finish this summary. Use the diagram you made on page 136.

_____

_____

In spring, the _____

in the mountains of Pakistan. The Hunza farmers

_____

_____ to bring

_____

_____

this water to their land. This water _____

_____

_____ .

Fall and Winter

Use your book to help you fill in the blanks.

What happens in fall?

_ _ _ _ _ _ _ _ _ _ _ _

1. The season after summer is _____ .

_ _ _ _ _ _ _ _ _ _

2. There are _____ hours of sunlight in fall than in summer.

3. Less sunlight makes the temperature

_ _ _ _ _ _ _ _ _ _ _ _

_____ .

_ _ _ _ _ _ _ _ _ _

4. Some _____ change color and fall off trees.

5. In fall, some animals store food or move to

_ _ _ _ _ _ _ _ _ _ _ _

_____ places.

What happens in winter?

6. The coldest season of the year is

_____

– – – – – – – –

_____ .

_____

– – – – – – – –

7. There are fewer hours of _____

in winter.

8. In winter there is not as much food for

_____

– – – – – – – –

_____ .

Critical Thinking

9. Why do you think some animals sleep until

spring?

_____

– – – – – – – – – – – – – –

_____

_____

_____

Name _____

Fall and Winter

Write the word *fall* or *winter* beside each picture to tell which season it shows.

1.

_____

-- -- -- -- -- -- -- -- -- --

_____

2.

_____

-- -- -- -- -- -- -- -- -- --

_____

3.

_____

-- -- -- -- -- -- -- -- -- --

_____

4.

_____

-- -- -- -- -- -- -- -- -- --

_____

5.

_____

-- -- -- -- -- -- -- -- -- --

_____

Fall and Winter

Fill in the blanks. Use the words from the box.

food	leaves	sunlight	winter

Fall and winter are the two coolest seasons of the year. In fall, there are fewer

_____

_ _ _ _ _ _ _ _

hours of _____ . Some

_____

_ _ _ _ _ _ _ _ _

trees lose their _____ ,

and many fruits get ripe.

The coldest season of the year is

_____

_ _ _ _ _ _ _ _ _ _

_____ . There is not

_____

_ _ _ _ _ _ _ _

enough _____ for

animals to eat. Some animals go to sleep or

leave for warmer places.

Seasons Change

✏️ Write About It

Write about one of the pictures. Describe the
weather. What could you wear and do
if you were there?

Getting Ready

Pick one of the pictures. Imagine yourself
there. What would you see, hear, and feel?
Write your ideas in the chart.

See	Hear	Feel

Drafting

Write your paragraph. Start with the main idea.
Describe the weather. What could you wear and
do if you were there?

_____

_ _

_____

_ _

_____

_ _

_____

_ _

_____

_ _

Now look at your paragraph.
Ask yourself:

▶ Did I begin with a main idea?

▶ Did I describe the weather?

▶ Did I correct all mistakes?

Weather and Seasons

Circle the words in each box that tell about the word at the top of the box.

1. seasons	2. weather tools
fall	wind vane
winter	rain gauge
temperature	clouds
summer	thermometer
spring	
3. water cycle	**4. temperature**
water vapor	cold
clouds	hot
rain	rain gauge
weather	thermometer

Draw a line from the picture to the word that tells about the picture.

1.    clouds

2.    wind vane

3.    fall

4.    rain gauge

5.    thermometer

6.    winter

Name _____

The Sky

Complete this diagram to show Earth and its neighbors. Use the words in the box.

Earth	Moon	planet	Sun

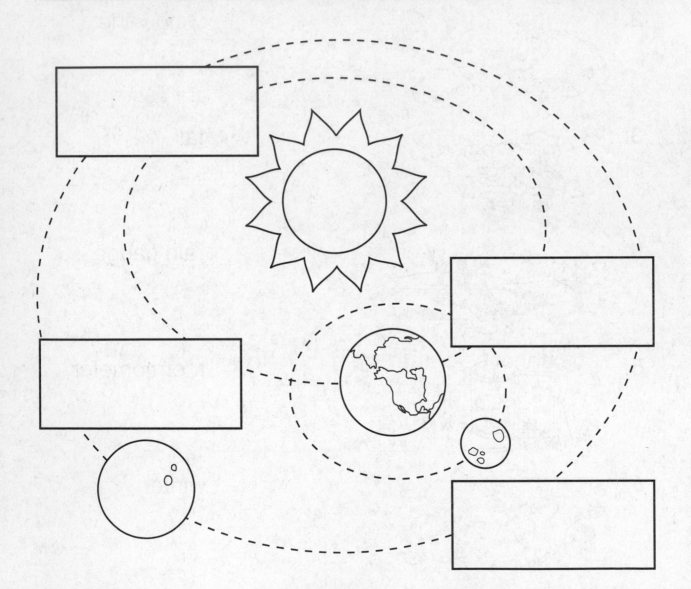

The Sky Above

Use your book to help you fill in the blanks.

What is in the sky?

_____

– – – – – – – – – –

1. In the dark _____ sky

   you might see clouds, the Moon, and stars.

   _____

   – – – – – – – – – –

2. The _____ are objects in

   the sky that make their own light.

   _____

   – – – – – – – – – –

3. The _____ is the star

   closest to Earth.

4. You cannot see other stars in the daytime,

   _____

   – – – – – – – – – –

   because the _____ is so

   bright, but they are still there.

Name _____

Why is the Sun important?

_____

_ _ _ _ _ _ _ _ _ _ _ _ _ _ _

5. The Sun makes _____ in

the form of heat and light.

_____

_ _ _ _ _ _ _ _ _ _ _ _ _

6. The Sun's energy _____

the air, land, and water, and helps plants grow.

7. Without the Sun, Earth would be too

_____

_ _ _ _ _ _ _ _ _ _ _ _ _

_____ and cold for living

things.

Critical Thinking

8. Would you plant a garden in the open or

under trees? Why?

_____

_ _ _ _ _ _ _ _ _ _ _ _ _ _ _ _

_____

_____

_ _ _ _ _ _ _ _ _ _ _ _ _ _ _ _

_____

The Sky Above

**Write what each picture shows about the sky.
Use each word in the box at least once.**

clouds	Moon	sky	stars	Sun

1.  _____
_ _ _ _ _ _ _ _ _ _ _ _ _ _ _
_____
_ _ _ _ _ _ _ _ _ _ _ _ _ _ _

2.  _____
_ _ _ _ _ _ _ _ _ _ _ _ _ _ _
_____
_ _ _ _ _ _ _ _ _ _ _ _ _ _ _

3.  _____
_ _ _ _ _ _ _ _ _ _ _ _ _ _ _
_____
_ _ _ _ _ _ _ _ _ _ _ _ _ _ _

Name _____

The Sky Above

Fill in the blanks. Use the words from the box.

heat	plants	stars	Sun

The Sun is a star that is closest to

Earth. The light of the

_____

_____ hides other stars

during the day. You can see many of these

_____

_____ in the night sky.

All living things depend on the Sun's

_____

_____ and light. Sunlight

_____

helps _____ grow.

Without the Sun, living things could not live

on Earth.

Earth Moves

Use your book to help you fill in the blanks.

What causes day and night?

_____

_ _ _ _ _ _ _ _ _

1. Earth _____ , or spins,

   very fast.

   _____

   _ _ _ _ _ _ _ _ _ _

2. When _____ rotates, we

   have day and night.

3. Day is when the place you are on Earth is

   _____

   _ _ _ _ _ _ _ _ _

   _____ the Sun.

4. Night is when Earth has rotated your place

   _____

   _ _ _ _ _ _ _ _

   _____ from the Sun.

   _____

   _ _ _ _ _ _ _ _ _

5. Shadows are _____

   when the Sun looks high in the sky.

Name _____

What causes a year?

6. While Earth spins each day, it also moves

_____

_ _ _ _ _ _ _ _ _ _ _ _

_____ the Sun.

_____

_ _ _ _ _ _ _ _ _ _ _

7. It takes one _____ for

Earth to make a full trip around the Sun.

8. As Earth travels around the Sun, the

_____

_ _ _ _ _ _ _ _ _ _ _

_____ change.

Critical Thinking

9. About how many times has Earth circled the

Sun since you were born?

_____

_ _ _ _ _ _ _ _ _ _ _ _ _ _ _

_____

_____

_ _ _ _ _ _ _ _ _ _ _ _ _ _ _

_____

_____

_ _ _ _ _ _ _ _ _ _ _ _ _ _ _

_____

Earth Moves

Tell how each person or thing rotates.

1.

_____

_____

_____

_____

2.

_____

_____

_____

_____

3.

_____

_____

_____

_____

Earth Moves

Fill in the blanks. Use the words from the box.

day	rotates	seasons

Many people once believed that the Sun moved around Earth. Today, we know that it is Earth that _____

_____, or spins, very fast as it travels around the Sun. This spinning _____

causes _____ and night.

Earth's trip around the Sun causes _____

the _____ to change. We cannot feel Earth move, but changes in seasons, and day and night show that it is moving very fast.

© Macmillan/McGraw-Hill

Time of Day

Use words that describe the sky.

✏ Write About It

About what time of day do you
think this picture was taken? Write
about things you can do at that time of day.

Getting Ideas

**Look at the picture. Think about the time of
day. Fill in the web below. Write the time of
day in the center. How do you know? Fill in
the other circles.**

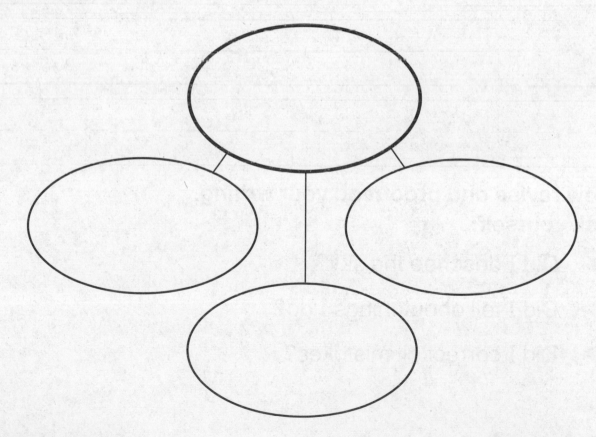

Drafting

Begin your story. Write a sentence that tells the time of day.

_____

– – – – – – – – – – – – – – – – – – –

_____

– – – – – – – – – – – – – – – – – – –

_____

Now write your story. Begin with the sentence you wrote. Describe the sky. Tell what things you do at that time of day.

_____

– – – – – – – – – – – – – – – – – – –

_____

– – – – – – – – – – – – – – – – – – –

_____

– – – – – – – – – – – – – – – – – – –

_____

Now revise and proofread your writing. Ask yourself:

▶ Did I describe the sky?

▶ Did I tell about things I do?

▶ Did I correct all mistakes?

Earth's Neighbors

Use your book to help you fill in the blanks.

How does the Moon look?

1. The Moon looks bright at night because the

   _____

   _ _ _ _ _ _ _ _ _

   _____ light shines on it.

2. While the Earth moves around the Sun, the

   _____

   _ _ _ _ _ _ _ _ _

   _____ moves around

   Earth.

   _____

   _ _ _ _ _ _ _ _ _ _

3. It takes the Moon about one _____

   to go once around Earth.

4. The shapes of the Moon that we see from

   _____

   _ _ _ _ _ _ _ _ _

   Earth are called _____ .

   _____

   _ _ _ _ _ _ _ _

5. You see the _____

   phases of the Moon each month.

What are planets?

_____

— — — — — — — — —

6. A _____ is a very large

object that moves around the Sun.

_____

— — — — — — — —

7. Earth is one of _____

planets moving around the Sun.

8. Some planets are smaller and others are

_____

— — — — — — — —

_____ than Earth.

_____

— — — — — — — — —

9. The planets closer to the _____

are warmer than those farther away.

Critical Thinking

10. In what ways are the planets different from

the Moon?

_____

— — — — — — — — — — — — — — — —

_____

Earth's Neighbors

Read the sentences. Write YES if the sentence is true. Write NO if the sentence is not true.

_____
– – – – – – – –
_____

1. The **Moon** travels around Earth.

_____
– – – – – – – –
_____

2. The **Moon** makes it own light.

_____
– – – – – – – –
_____

3. The Moon's **phases** are caused by clouds.

_____
– – – – – – – –
_____

4. The same pattern of Moon **phases** happens each month.

_____
– – – – – – – –
_____

5. There are eight **planets** moving around our Sun.

_____
– – – – – – – –
_____

6. Earth is the only **planet** with living things.

Name _____

Earth's Neighbors

Fill in the blanks. Use the words from the box.

Moon	phases	planets

When you look up at the night sky,
what do you see? If the night is clear, you
_____

_ _ _ _ _ _ _ _ _

may see the _____ ,
which is Earth's closest neighbor.
Each night in a month, the Moon's
_____

_ _ _ _ _ _ _ _ _

_____ change.

_____

_ _ _ _ _ _ _ _ _

You may see _____ ,
like Earth, in the sky. Earth is the only planet
that has living things.

Use with **Lesson 3**
Earth's Neighbors

Meet Ben Oppenheimer

Read the Reading in Science pages in your book. Look for a cause and its effect as you read. Remember, a cause is why an event happens. An effect is an event that happens. Fill in the diagram below. Tell what causes scientists to use telescopes.

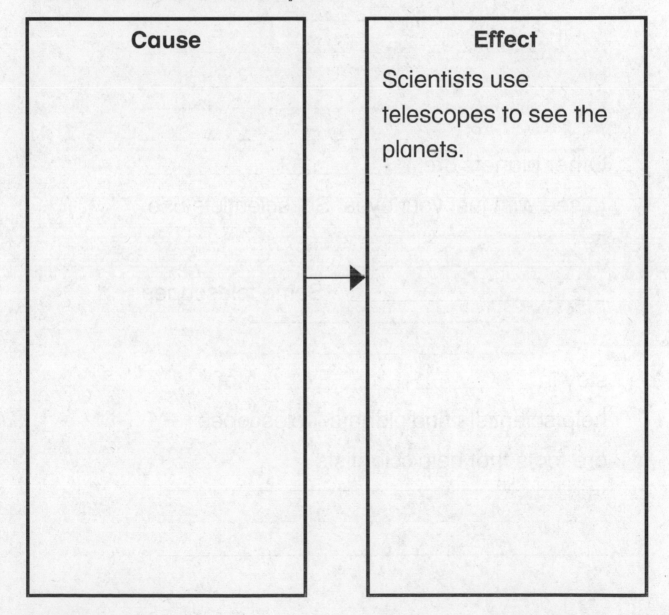

Cause	Effect
	Scientists use telescopes to see the planets.

✏ Write About It

Cause and Effect. Why do scientists use telescopes? Finish this explanation. Use the information in the Cause and Effect Chart you made to help you.

You can see some planets, like Venus,

_____

_ _ _ _ _ _ _ _ _ _ _ _ _

by _____ .

_____

_ _ _ _ _ _ _ _ _ _

Other planets are _____

to see with just your eyes. So, scientists use

_____

_ _ _ _ _ _ _ _ _ _

_____ . Some telescopes

_____

_ _ _ _ _ _ _ _ _

stay in _____ to

help scientists find planets. Telescopes

are tools that help scientists

_____

_ _ _ _ _ _ _ _ _ _ _ _ _

_____ .

The Sky

**Use the words in the box and the clues below
to answer each riddle.**

Moon	phases	planets

I. Sometimes they are sliver thin.

Sometimes full and round.

We can see these monthly changes,

Watching from the ground.

What are they?

_____   _____

— — — — — — — — — —   — — — — — — — — —

_____ of the _____

2. There are eight of these objects

Of different sizes and shapes.

Around and around the Sun they go,

A regular path they take.

What are they?

_____

— — — — — — — — —

_____

Unscramble each word. Write it on the line to answer each question.

3. sart   What is the Sun? _____

_ _ _ _ _ _ _ _ _ _

4. tlnpae   What is Earth? _____

5. noMo   What shines with the Sun's light?

_____

_ _ _ _ _ _ _ _ _ _

_____

6. uSn   What does Earth travel around?

_____

_ _ _ _ _ _ _ _ _ _

_____

7. hsspae   What are the Moon's changes as seen

_____

_ _ _ _ _ _ _ _ _ _

from Earth? _____

Where in the World Is Water?

Read the Unit Literature pages in your book.

National Wildlife Federation
Ranger Rick

✏ **Write About It**

Response to Literature

1. How does the article show the way water

turns into a gas?

_____

– – – – – – – – – – – – – – – – –

_____

– – – – – – – – – – – – – – – – –

_____

2. The article says that water covers most of

Earth. Which form of water do you think

covers most of Earth's surface? Why?

_____

– – – – – – – – – – – – – – – – –

_____

– – – – – – – – – – – – – – – – –

_____

– – – – – – – – – – – – – – – – –

_____

Matter Everywhere

The name of each state of matter is next to each box. As you read the chapter, write at least two properties for each state of matter.

States of Matter and Their Properties	
Solids	
Liquids	
Gases	

Describing Matter

Use your book to help you fill in the blanks.

What is matter?

I. When you describe something, you talk
_____

_ _ _ _ _ _ _ _ _

about its _____ .

2. Some properties of matter are color, size,
_____

_ _ _ _ _ _ _ _ _

and _____ .

_____

_ _ _ _ _ _ _ _ _ _

3. All things are made of _____ .

_____

_ _ _ _ _ _ _ _

4. Matter comes in _____

forms.

_____

_ _ _ _ _ _ _ _ _

5. Solid, _____ , and liquid

are forms of matter.

_____

_ _ _ _ _ _ _ _

6. Matter takes up _____ .

Name _____

What is mass?

7. How much matter is in an object is its

   _____

   _ _ _ _ _ _ _ _ _

   _____ .

8. Heavier objects have _____

   mass than lighter objects.

9. You can measure mass with a

   _____

   _ _ _ _ _ _ _ _ _

   _____ .

Critical Thinking

10. What is matter? How can you describe it?

   _____

   _ _ _ _ _ _ _ _ _ _ _ _ _ _ _

   _____

   _____

   _ _ _ _ _ _ _ _ _ _ _ _ _ _ _

   _____

   _ _ _ _ _ _ _ _ _ _ _ _ _ _ _

   _____

Describing Matter

Circle the words in each box that tell about the word at the top of the box.

---

1. properties

color	smell
block	taste
size	car
shape	mass

---

2. matter

ten

gas

liquid

solid

3. mass

balance

empty

lighter

heavier

Name _____

Describing Matter

Fill in the blanks. Use the words from the box.

| mass | smell | solid | taste |

Every kind of matter has its own special properties. _____

_ _ _ _ _ _ _ _ _ _

Think about the color, _____ ,

and even the sound of popping corn.

Now think about its _____

_ _ _ _ _ _ _ _ _

_____ and its special

shape. Each is a property of the _____

_ _ _ _ _ _ _ _ _

_____ we call popcorn.

And let's not forget popcorn's delicious _____

_ _ _ _ _ _ _ _

_____ . Many people

think it's the best property of this kind of

matter!

A Shoe Story

✏ Write About It

Look at your shoes. What are the properties
of your shoes? Draw and write about your
shoes. What do your shoes tell about you?

Getting Ready

Look closely at your shoes. Answer Who?
What? When? Where? and How? questions
about them.

Who is the description about?

What is special about them?

Where do you wear them?

When do you wear them?

How do you feel when you wear them?

Drafting

The main idea is the most important idea.
Write a main idea for your paragraph.

_____

_ _

_____

_ _

Now write your paragraph. Begin with your
main idea sentence. Make sure you have a
clear beginning, middle, and end to your story.

_____

_ _

_____

_ _

_____

_ _

_____

Now look at your paragraph. Ask yourself:

▶ Did I describe the properties of my shoes?

▶ Did I tell what my shoes show about me?

© Macmillan/McGraw-Hill

Solids

Use your book to help you fill in the blanks.

What is a solid?

_____

_ _ _ _ _ _ _ _ _ _

1. The amount of _____ in a

   solid stays the same.

   _____

   _ _ _ _ _ _ _ _ _

2. A _____ is a form of

   matter.

   _____

   _ _ _ _ _ _ _ _ _

3. Only a solid has a _____

   of its own.

   _____

   _ _ _ _ _ _ _ _ _

4. A solid will keep the _____

   shape when it is moved.

What are some properties of solids?

5. Solids come in different sizes, shapes, and

   _____

   _ _ _ _ _ _ _ _ _

   _____ .

_____

6. You can fold or _____

some solids.

7. Solids feel different because they have

_____

_ _ _ _ _ _ _ _ _ _ _ _ _

different _____ .

_____

_ _ _ _ _ _ _ _ _ _ _ _

8. You can use a _____ to

measure the length of some solids.

Critical Thinking

9. If you rolled one ball of clay into 4 smaller

balls, what would happen to its mass? How

could you check?

_____

_ _ _ _ _ _ _ _ _ _ _ _ _ _ _ _ _ _ _

_____

_ _ _ _ _ _ _ _ _ _ _ _ _ _ _ _ _ _ _

_____

Solids

Look at the pictures. Color each object that is a solid.

Name _____

Solids

Fill in the blanks. Use the words from the box.

matter	same	shape	solid

Blocks, balls, and dolls are all
toys. Each of these objects is also a

_____

_ _ _ _ _ _ _ _ _

_____ . A solid is one

_____

_ _ _ _ _ _ _ _ _

form of _____ .

A solid does not change its

_____

_ _ _ _ _ _ _ _ _

_____ even when you

move it. The total amount of matter in a solid

_____

_ _ _ _ _ _ _ _ _ _

always stays the _____ .

All these toys are alike because they are the

same state of matter.

© Macmillan/McGraw-Hill

Building Blocks

Read the Reading in Science pages in your book. Look for information that can help you make a prediction as you read. Fill in the chart below. Write a prediction about which pig's house would last the longest.

What I Predict	Clues I Used

Name _____

Write About It

Predict. What would a straw house look
like after a hundred years? You may use the
information in the Prediction Chart you made
to help you.

I predict that a straw house would be

_____

_ _ _ _ _ _ _ _ _ _ _ _ _ _ _ _ _ _

_____ in a hundred years.

_____

_ _ _ _ _ _ _ _ _ _ _ _ _ _ _ _ _ _

I think this because _____

_____

_ _ _ _ _ _ _ _ _ _ _ _ _ _ _ _ _ _

_____

_____

_ _ _ _ _ _ _ _ _ _ _ _ _ _ _ _ _ _

_____

Liquids and Gases

Use your book to help you fill in the blanks.

What are some properties of liquids?

1. A _____ is a state
   of matter.

2. Liquids have _____
   and take up space.

3. Liquids do _____ have
   a shape of their own.

4. Liquids take the _____
   of whatever they are in.

5. You can use a _____
   to measure liquids.

What are some properties of gases?

_____

6. A third form of matter is a _____ .

_____

7. Gases do not have their own _____ .

_____

8. Gases spread out _____ to fill space.

_____

9. The _____ we breathe is made up of gases.

Critical Thinking

10. How are a liquid and a gas alike and different?

_____

_____

_____

_____

Liquids and Gases

Read each sentence. If it is true, check the YES column. If it is not true, check the NO column.

	YES	NO
1. A liquid's shape will look different in different containers.		
2. Liquids always have less mass than solids.		
3. Liquids flow in different ways.		
4. Air is made up of different gases.		
5. Gases always spread out evenly to fill whatever they are in.		
6. A measuring cup can be used to measure a gas.		

Liquids and Gases

Fill in the blanks. Use the words from the box.

flows	gas	liquid	shape

The glass you pour milk into is
a solid. But the milk you fill it with is a

_____

_ _ _ _ _ _ _ _ _ _

_____ . When you pour

_____

_ _ _ _ _ _ _ _ _ _

the milk, it _____ into

the glass. The milk takes the

_____

_ _ _ _ _ _ _ _ _ _

_____ of the glass.

_____

_ _ _ _ _ _ _ _ _ _

A _____ also takes

the shape of what you put it in. When you

blow into a balloon, the air spreads to fill all

the space.

© Macmillan/McGraw-Hill

Matter Everywhere

Fill in the blanks with the words from the box.

balance	mass	matter	properties

_____

1. _____ how much matter

is in something

_____

2. _____ how something

looks, smells, feels, or tastes

_____

3. _____ something used to

measure mass

_____

4. _____ what all things are

made of

**Write the word that goes with each meaning.
Use the boxed letters.**

1. the amount of matter in something

___ ___ [s] ___

2. the color, size, and shape of something

___ ___ [o] ___ ___ ___ ___ ___

3. a kind of matter that takes the shape of
what it's in

[l] ___ ___ ___ [i] [d]

**What's hard or soft, or big or small, but will
keep its shape even when it is moved? Use
the boxed letters.**

A [s] [ ] [ ] [ ] [d]

© Macmillan/McGraw-Hill

Changes in Matter

Tell how matter can change. Use the words in the box.

gas	liquid	solid

Heat a solid.	→	_____

Cool a liquid.	→	_____

Heat a liquid.	→	_____

Name _____

Matter Can Change

Use your book to help you fill in the blanks.

How can matter change?

_____

– – – – – – – – –

1. You can fold, bend, or tear a _____ to change it.

_____

– – – – – – – – –

2. The solid will have a _____ shape, but it is still the same kind of matter.

_____

– – – – – – – – –

3. You can sometimes _____ matter into something else.

_____

– – – – – – – – –

4. The _____ of matter can be changed by heat and air.

5. When matter _____

it changes into something different.

6. When paper burns, it turns into

_____

_ _ _ _ _ _ _ _ _ _ _

_____ .

_____

_ _ _ _ _ _ _ _ _

7. It is no longer _____ .

Critical Thinking

8. What are some ways to change a solid

without changing its matter? How can you

change a solid by changing its matter?

_____

_ _ _ _ _ _ _ _ _ _ _ _ _

_____

_ _ _ _ _ _ _ _ _ _ _ _ _

_____

_____

Name _____

Matter Can Change

Look at these pictures. Color the one that shows how burning can change matter. Then finish the sentence about this picture.

_____

When you burn _____ it

changes _____

Matter Can Change

Fill in the blanks. Use the words from the box.

bend	fire	looks	paint

You can change a solid like a
cardboard tube in different ways. You can

_____

_ _ _ _ _ _ _ _ _ _ _

tear or _____ it to make a

_____

_ _ _ _ _ _ _ _ _

toy animal. You can _____

it to make a face. These things change how

_____

_ _ _ _ _ _ _ _ _

the tube _____ , but the

tube is still cardboard.

If you put a cardboard tube in a

_____

_ _ _ _ _ _ _ _ _

_____ , it will burn.

The matter is changed.

Making Mixtures

Use your book to help you fill in the blanks.

What is a mixture?

_____

_ _ _ _ _ _ _ _

1. A _____ is two or more

   things put together.

   _____

   _ _ _ _ _ _ _ _ _

2. When you mix _____ ,

   they do not change.

   _____

   _ _ _ _ _ _ _ _ _

3. You can pick _____

   the solids in a mixture.

   _____

   _ _ _ _ _ _ _ _ _

4. Some solids _____

   in water and others sink.

   _____

   _ _ _ _ _ _ _ _

5. You can sometimes _____

   solids out of water.

What are some other mixtures?

_____

6. Some solids can _____

into a liquid.

7. When you mix water with another

_____

_ _ _ _ _ _ _ _ _

_____ , it may mix

completely.

_____

_ _ _ _ _ _ _ _

8. Some liquids, like _____ ,

do not mix completely.

Critical Thinking

9. When was the last time you made a mixture?

What was in it? What happened to the

different parts?

_____

_ _ _ _ _ _ _ _ _ _

_____

_____

_ _ _ _ _ _ _ _ _ _

Making Mixtures

Complete the sentence that tells about each picture. Use the words in the box.

dissolve	mixture

1. Anna is using a

_____

_ _ _ _ _ _ _ _ _ _ _ _

_____

of round and square beads
to make a necklace.

2. Tim stirs to help the solid

_____

_ _ _ _ _ _ _ _ _ _ _ _

_____

in water when he makes
juice.

© Macmillan/McGraw-Hill

Making Mixtures

Fill in the blanks. Use the words from the box.

dissolve	mixture	separate	solids

Have you ever made trail mix? This is an

_____

easy _____ to make. Put

_____

some _____ like raisins,

nuts, and cereal in a bowl. Then stir everything

well. You can pick out the different solids if you

want.

Lemonade is a mixture you cannot

_____

_____ . Put lemon juice,

sugar, and water in a pitcher. Then stir to

_____

_____ the sugar. This is a

mixture that mixes completely with the water.

Mix It Up

✏️ **Write About It**

Write a story about the picture.

Tell about the mixture in the bag.

Can you take it apart?

Getting Ready

Write all the things you see in the backpack in the circles.

Drafting

Write a sentence to begin your story about the mixture. Then write your story. Use words that tell how something looks.

_____

– –

_____
_____

– –

_____

– –

_____

Now look at your paragraph. Ask yourself:

▶ Did I describe the things in the backpack?

▶ Did I correct all mistakes?

Name _____

Heat Can Change Matter

Use your book to help you fill in the blanks.

How can solids and liquids change?

_____

_ _ _ _ _ _ _ _ _

1. A liquid gets _____ when it loses heat.

_____

_ _ _ _ _ _ _ _ _ _

2. Liquids _____ when they get very cold.

_____

_ _ _ _ _ _ _ _

3. A liquid becomes a _____ when it freezes.

_____

_ _ _ _ _ _ _ _ _

4. When you _____ a solid, you change it into a liquid.

5. Some solids can melt with a little
_____

_ _ _ _ _ _ _ _

_____ , while others need a lot.

How can liquids and gases change?

6. When water is heated, some water

_____

– – – – – – – – – –

_____ .

_____

– – – – – – – – – – –

7. The more _____ added

to water, the faster it will evaporate.

_____

– – – – – – – – – –

8. Water as a gas is called _____ .

_____

– – – – – – – –

9. When water vapor is _____ ,

it turns back into a liquid.

Critical Thinking

10. What happens to water when it is cooled and

heated?

_____

– – – – – – – – – – – – – –

_____

_____

– – – – – – – – – – – – – –

_____

Heat Can Change Matter

Solve these riddles. Use the words in the box.

evaporate	melt

1. A snowman is a solid

as cute as he can be.

But the sun's heat will change him,

before the clock strikes three!

_____

_ _ _ _ _ _ _ _

The snowman will _____ .

2. A puddle in the farmyard,

is fun for a duck or a hen.

But the heat will make it vanish,

by no later than ten!

_____

_ _ _ _ _ _ _ _

The water will _____ .

Heat Can Change Matter

Fill in the blanks. Use the words from the box.

evaporate	freeze	melt

On Saturday, Kim left her empty pail
in the sandbox. On Sunday it rained. Water
filled the pail. On Monday it was cold. The

_____

— — — — — — — — —

water began to _____

into ice.

On Wednesday the Sun came out. The

_____

— — — — — — — — —

ice began to _____ into

liquid. On Thursday and Friday,

the Sun's heat made the water

_____

— — — — — — — — —

_____ . By Saturday, Kim

found her empty pail where she had left it.

Name _____

Hot Stuff

Read the Reading in Science pages in your book. Look for problems and their solutions as you read. Fill in the chart below. Write a solution for the problem.

Problem	Solution
Your hot chocolate is losing heat too fast.	_____ _____ _____ _____ _____

© Macmillan/McGraw-Hill

✏ Write About It

Problem and Solution. What could you do if you wanted to keep apple cider warm? Finish this solution using what you have read.

If you wanted to keep apple cider warm, you

_____

_ _

could pour it into a _____ .

_____

_ _

This is because heat travels _____

_____

_ _

_____ .

Then you could put a lid on it to keep the heat

_____

_ _

_____ .

Changes in Matter

Find the six vocabulary words hiding in the puzzle. Then circle them. Look across and down.

burn	evaporate	melt
dissolve	freeze	mixture

m	e	l	l	d	s	t	i	p
p	w	e	r	i	s	t	m	e
n	l	i	q	s	b	d	i	b
m	a	r	s	s	u	t	x	m
e	v	a	p	o	r	a	t	e
l	k	l	o	l	n	i	u	x
t	a	r	p	v	e	x	r	t
m	e	f	r	e	e	z	e	t

Name _____

Draw a line from the picture to the word that tells about the picture.

1.    dissolve

2.    freeze

3.    burn

4. 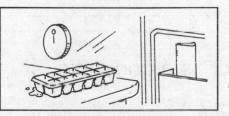   mixture

5. 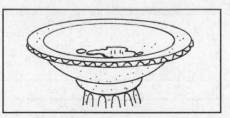   melt

6.    evaporate

For A Quick Exit

by Norma Farber

Read the Unit Literature pages in your book.

✏ Write About It

Response to Literature

1. Why does the poet say that escalators come
   in pairs?

   _____

   _____

   _____

   _____

2. Do you think "For A Quick Exit" is a good title
   for the poem? Why or why not?

   _____

   _____

   _____

   _____

3. Draw a picture to show what the poet is
   writing about. Use a separate sheet of paper.

On the Move

Fill in the important ideas as you read the chapter. Use the words in the box. Some words may fit in more than one place.

attract	lever	pulley	ramp
gravity	pull	push	repel

On the Move		
What forces make things move?	**What simple machines move things?**	**How do magnets move things?**

Name _____

Position and Motion

Use your book to help you fill in the blanks.

How can you tell where something is?

1. An object's _____ is the

   place where it is located.

2. Words such as above, below, left, and

   _____

   _____ can describe an

   object's position.

How do things move?

3. The word _____

   describes a change in an object's position.

4. Objects can move in a _____

   or a curvy line.

5. They can move _____ or down.

© Macmillan/McGraw-Hill

6. Objects can move around in a

_____

_ _ _ _ _ _ _ _ _ _ _

_____ or can zigzag.

_____

_ _ _ _ _ _ _ _ _ _

7. Speed is how fast or _____

an object moves.

_____

_ _ _ _ _ _ _ _ _ _ _

8. Some objects move at a fast _____

while other objects move more slowly.

Critical Thinking

9. How do you move on a swing, on a seesaw,

or on a merry-go-round?

_____

_ _ _ _ _ _ _ _ _ _ _ _

_____

_____

_ _ _ _ _ _ _ _ _ _ _

_____

_____

_ _ _ _ _ _ _ _ _ _ _

_____

Position and Motion

Complete the sentence that tells about each picture. Use the words in the box.

motion	position	speed

I.

The bike rider moves at a faster

_____

– – – – – – – – – – –

_____ than the

skater.

2.

_____

– – – – – – – – – – –

The cat's _____

is above the puppy.

3.

The butterflies fly in a zigzag

_____

– – – – – – – – – – –

_____ from

flower to flower.

Position and Motion

Fill in the blanks. Use the words from the box.

motion	right
position	speeds

Have you ever watched a race?
Runners begin by taking their

_____

_____ at the starting

line. The runners are on the

_____

_____ and left along

the line. At the signal, the runners

_____

_____

go into _____ .
Different runners move at different

_____

_____ down the track.
The fastest runner is the winner!

© Macmillan/McGraw-Hill

Name _____

Pushes and Pulls

Use your book to help you fill in the blanks.

What makes things move?
_____
_ _ _ _ _ _ _ _ _ _ _

1. A _____ is what makes
   things move.

   _____
   _ _ _ _ _ _ _ _ _ _ _

2. A force can be a _____
   or a pull that makes things start moving.

   _____
   _ _ _ _ _ _ _ _ _ _ _

3. A push moves something _____
   from you.

   _____
   _ _ _ _ _ _ _ _ _ _ _

4. A _____ moves something
   closer to you.

   _____
   _ _ _ _ _ _ _ _ _ _ _

5. A force called _____ pulls
   things toward Earth.

How are forces different?

6. How things move depends on the

_____

_ _ _ _ _ _ _ _

_____ of the push or pull.

7. A larger force can make an object move faster

_____

_ _ _ _ _ _ _ _

and farther than a _____

force.

_____

_ _ _ _ _ _ _ _

8. A force called _____

makes things slow down.

9. Friction happens when two objects

_____

_ _ _ _ _ _ _ _

_____ together.

Critical Thinking

10. Why do many sneakers have rough bottoms?

_____

_ _ _ _ _ _ _ _ _ _ _ _

_____

_____

_ _ _ _ _ _ _ _ _ _ _ _

_____

Pushes and Pulls

Circle the word that tells about the force shown in the picture.

1.

pull   force   gravity

2.

gravity   push   pull

3.

friction   gravity   force

4.

push   friction   pull

Pushes and Pulls

Fill in the blanks. Use the words from the box.

force	friction	pull	push

Suppose you want to move your

toy box across the room. It will take

_____

– – – – – – – – – –

a _____ to make

it move. You can stand behind it and

_____

– – – – – – – – – –

_____ it. You can stand in

_____

– – – – – – – – – –

front of it and _____ on

its handle.

If the box is on a rough carpet,

_____

– – – – – – – – – –

_____ will make it hard

to move. The friction will slow it down.

Simple Machines

Use your book to help you fill in the blanks.

What are simple machines?

_____

_ _ _ _ _ _ _ _ _

1. A _____ makes it easier

   to move things and do work.

2. A rope that moves over a wheel is called a

   _____

   _ _ _ _ _ _ _ _ _

   _____ .

   _____

   _ _ _ _ _ _ _ _ _

3. A pulley makes it easier to _____

   things.

   _____

   _ _ _ _ _ _ _ _ _

4. Pulleys also help _____

   things, like flags, to high places.

What are levers and ramps?

5. A ramp makes it easier to move things

   _____

   _ _ _ _ _ _ _ _ _

   _____ or down.

_____

_ _ _ _ _ _ _ _

6. A _____ is a bar that

balances on a fixed point.

_____

_ _ _ _ _ _ _ _ _

7. A lever can help you _____

or move things.

_____

_ _ _ _ _ _ _ _ _

8. Seesaws and _____

are examples of levers.

Critical Thinking

9. How would you use a lever and a log to lift a

heavy rock?

_____

_ _ _ _ _ _ _ _ _ _ _ _ _ _

_____

_ _ _ _ _ _ _ _ _ _ _ _ _ _

_____

_____

Name _____

Simple Machines

Circle the name of each simple machine shown in the picture. Then tell how people use it.

1. ramp pulley lever	
2. lever pulley ramp	
3. pulley lever ramp	

Use with **Lesson 3**
Simple Machines

Simple Machines

Fill in the blanks. Use the words from the box.

lever	pulleys	raise	ramps

Your school is filled with many kinds of simple machines. A seesaw on the playground

_____

is an example of a _____ .

_____

The _____ on the window shades, help you

_____

_____ the shade. Your

_____

school may have _____ by the outside doors to help people in wheelchairs. All of these simple machines make it easier for people to work and play.

Moving Up

Read the Reading in Science pages in your book. Think about how to classify information as you read. Remember, when you classify, you put things that are alike into groups. Fill in the chart below.

Where do elevators and escalators belong?

Things that Have Pulleys	Things that Have Ramps
_____ _____ _____	_____ _____ _____

How do pulleys and ramps help you everyday?

_____

_____

_____

_____

_____

© Macmillan/McGraw-Hill

✏ Write About It

Classify. What is something else that can fit into one of the groups from the Classify Chart you made? Draw it. Then tell which group it would fit in.

```
┌──────────────────────────────────────────┐
│                                          │
│                                          │
│                                          │
│                                          │
│                                          │
│                                          │
│                                          │
│                                          │
└──────────────────────────────────────────┘
```

A _____ would fit in the group

_____

Things that Have _____ .

Magnets

Use your book to help you fill in the blanks.

What is a magnet?

_____

_ _ _ _ _ _ _ _ _ _

1. A _____ can pull, or

   attract some objects.

2. Magnets attract objects that have

   _____

   _ _ _ _ _ _ _ _ _

   _____ in them.

   _____

   _ _ _ _ _ _ _ _ _ _

3. Iron is a type of _____

   found in many objects.

What are a magnet's poles?

4. Every magnet has a north and a south

   _____

   _ _ _ _ _ _ _ _ _ _

   _____ .

5. A magnet's poles are where the

_____

– – – – – – – – –

_____ is strongest.

6. The north pole of one magnet will

_____

– – – – – – – – –

_____ the south pole of

another magnet.

_____

– – – – – – – – –

7. Magnets will _____ , or

push each other apart, if you put two like

poles next to each other.

Critical Thinking

8. Why will a magnet pick up some paper clips

and not others?

_____

– – – – – – – – – – – – – – – – –

_____

_____

– – – – – – – – – – – – – – – – –

_____

Name _____

Magnets

Look at each picture. Tell if these magnets will attract or repel each other and explain why.

 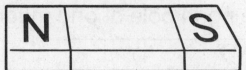

_____

_ _ _ _ _ _ _ _ _ _ _ _ _

1. These magnets will _____ each

other because _____

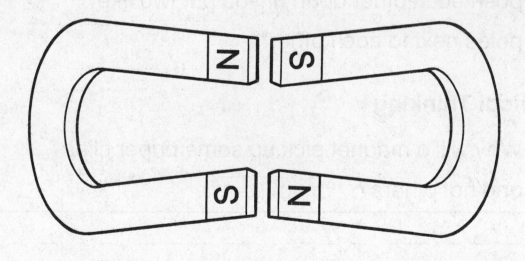

_____

_ _ _ _ _ _ _ _ _ _ _ _ _

2. These magnets will _____ each

other because _____

Magnets

Fill in the blanks. Use the words from the box.

attract	magnet	pole

Many toys are made with magnets. A toy

_____

_ _ _ _ _ _ _ _ _ _

train can have a _____ on

both ends of each car. When the north

pole of one magnet is near the south

_____

_ _ _ _ _ _ _ _ _ _

_____ of another magnet,

the cars will stick together.

A toy fishing rod with a magnet will

_____

_ _ _ _ _ _ _ _ _ _

_____ fish made with paper

clips on them. Magnets can help us do work

and also have fun!

Fun with Magnets

Tell what happens first, next, and last.

✏ Write About It

Explain how magnets are used in this picture. Write a story about how you use magnets.

Getting Ideas

Choose a toy that uses magnets. Fill in the chart below. Tell how magnets make the toy work.

First
↓

Next
↓

Last

© Macmillan/McGraw-Hill

Drafting

Begin your story. Look at the picture. Write a sentence. Tell how the girl used magnets.

_____

- -

_____

- -

_____

Now write a story about how you use magnets. Tell what happens first, next and last.

- -

_____

- -

_____

- -

_____

- -

_____

**Now revise and proofread your writing.
Ask yourself:**

▶ Did I tell what happens first, next, and last?

▶ Did I correct all mistakes?

On the Move

Circle the words that tell about each word at the top.

1. simple machines

   ramp   gravity   lever   pulley

2. position

   above   under   friction   left

3. magnet

   repel   attract   poles   position

4. motion

   speed   fast   zigzag   poles

Circle the twelve vocabulary words hiding in the puzzle. Look across and down.

force	gravity	magnet	pull	push	repel
friction	lever	poles	pulley	ramp	speed

```
t  f  o  r  c  e  l  g  o  r  t
e  r  c  e  s  f  x  r  a  m  p
s  i  f  p  u  s  h  a  s  a  u
y  c  p  e  i  p  e  v  m  g  l
k  t  u  l  a  e  m  i  v  n  l
n  i  l  e  v  e  r  t  i  e  e
p  o  l  e  s  d  r  y  s  t  y
b  n  c  u  m  p  w  v  g  x  l
```

Name _____

Energy Everywhere

Fill in the web below. What are the different kinds of energy? Use the words in the box.

| electricity | heat | light | sound |

Kinds of Energy

What makes things warm?

_____
- - - - - - - -
_____

What gives some things power to work?

_____
- - - - - - - -
_____

What is made when an object vibrates?

_____
- - - - - - - -
_____

What lets you see?

_____
- - - - - - - -
_____

Energy and Heat

Use your book to help you fill in the blanks.

What is energy?

_____

– – – – – – – – – – – –

1. Eating food gives you _____

   to work and play.

2. We need energy to make things

   _____

   – – – – – – – – – – –

   _____ and change.

   _____

   – – – – – – – – – – –

3. Electricity, heat, _____ ,

   and sound are forms of energy.

4. Gasoline gives cars

   _____

   – – – – – – – – – – –

   _____ to move.

5. Windmills turn energy from wind into

   _____

   – – – – – – – – – – –

   _____ .

What is heat?

_____

6. One form of energy is _____

   that makes things warm.

   _____

   — — — — — — — — — —

7. The _____ gives us most

   of the heat energy on Earth.

   _____

   — — — — — — — — —

8. Heat also comes from _____

   things like wood, oil, and gas.

   _____

   — — — — — — — — — —

9. You can also make heat by _____

   things together.

Critical Thinking

10. What are two ways that you used energy

    today?

    _____

    — — — — — — — — — — — — —

    _____

Energy and Heat

Tell how each person is getting or using heat energy.

1.

_ _ _ _ _ _ _ _ _ _ _ _ _ _ _ _ _
_____
_ _ _ _ _ _ _ _ _ _ _ _ _ _ _ _ _
_____
_ _ _ _ _ _ _ _ _ _ _ _ _ _ _ _ _

2.

_ _ _ _ _ _ _ _ _ _ _ _ _ _ _ _ _
_____
_____
_ _ _ _ _ _ _ _ _ _ _ _ _ _ _ _ _

3.

_____
_ _ _ _ _ _ _ _ _ _ _ _ _ _ _ _ _
_____
_ _ _ _ _ _ _ _ _ _ _ _ _ _ _ _ _

Energy and Heat

Fill in the blanks. Use the words from the box.

food	gas	heat

You use different forms of energy each

_____

— — — — — — — — —

day. The _____ you eat

gives you energy to work and play at school.

You may come to school by bus, that runs

_____

— — — — — — — — —

on _____ . If the day is

_____

— — — — — — — — —

cool, _____ energy from

burning oil may warm the room.

Almost everything you do uses some

form of energy.

© Macmillan/McGraw-Hill

Sound

Use your book to help you fill in the blanks.

How can you make sound?

_____

_ _ _ _ _ _ _ _ _ _ _

1. Sound is a form of _____

   that you can hear and sometimes feel.

2. Sound is made when something
   _____

   _ _ _ _ _ _ _ _ _

   _____ , or moves back

   and forth.

3. When an object stops vibrating, the
   _____

   _ _ _ _ _ _ _ _ _

   _____ stops too.

   _____

   _ _ _ _ _ _ _ _ _

4. Different objects make _____

   sounds.

   _____

   _ _ _ _ _ _ _ _ _

5. Some sounds _____ us

   things, like an alarm clock wakes you up.

How are sounds different?

_____

– – – – – – – – – –

6. Big vibrations make _____

sounds.

_____

– – – – – – – – – –

7. Small vibrations make _____

sounds.

_____

– – – – – – – – – –

8. Fast vibrations make _____

pitched sounds.

_____

– – – – – – – – – –

9. Slow vibrations make _____

pitched sounds.

Critical Thinking

10. How can you change the sound of a drum?

_____

– – – – – – – – – – – – – – – – – – – –

_____

– – – – – – – – – – – – – – – – – – – –

_____

Sound

Circle the pitch of each object.

1.

high pitch   low pitch

2.

high pitch   low pitch

3.

high pitch   low pitch

4.

high pitch   low pitch

Sound

Fill in the blanks. Use the words from the box.

feel	pitch	sound	vibrate

Sounds are all around you. Like

_____

_ _ _ _ _ _ _ _ _

heat, _____ is a

form of energy. Sounds are caused

_____

_ _ _ _ _ _ _ _ _

when things _____ .

You can only hear and sometimes

_____

_ _ _ _ _ _ _ _ _

_____ sound.

A sound can have a high

_____

_ _ _ _ _ _ _ _ _

_____ or a low pitch.

What makes your favorite sound?

Sounds and Safety

Read the Reading in Science pages in your book. Fill in the diagram below. Write the important ideas in the small boxes. Then retell these ideas in the big box.

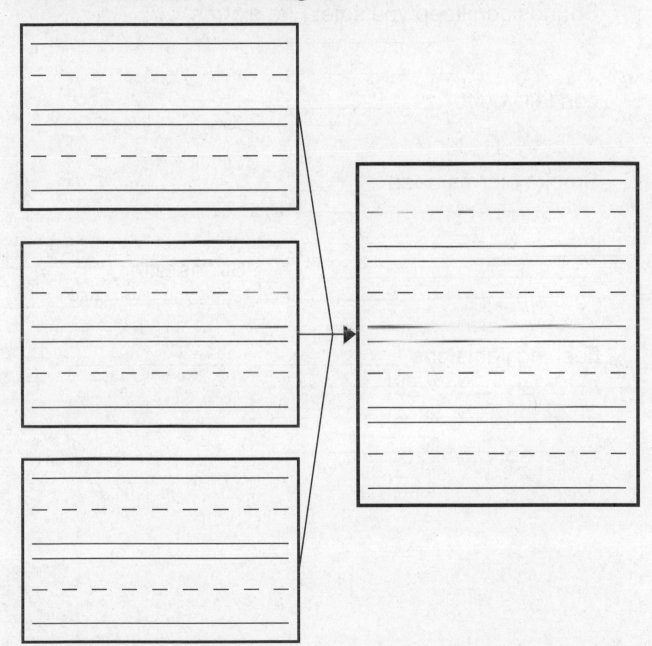

✎ Write About It

Summarize. How can sounds keep you safe? Finish this summary. Use the Summarize Chart you made.

Sounds can keep you safe. Fire alarms

_____

– – – – – – – – –

can tell you to _____ .

_____

– – – – – – – –

Smoke alarms beep _____

_____

– – – – – – – –

_____ . Sirens with

_____

– – – – – – – – –

flashing lights can _____

_____

– – – – – – – –

_____ .

Light

Use your book to help you fill in the blanks.

What is light?

_____

_ _ _ _ _ _ _ _ _ _ _ _ _

1. Light is a kind of _____

    that lets you see.

2. Different objects allow different amounts of

    _____

    _ _ _ _ _ _ _ _ _ _ _

    _____ to pass through

    them.

3. Some objects do not allow any light to

    _____

    _ _ _ _ _ _ _ _ _ _

    _____ through them.

    _____

    _ _ _ _ _ _ _ _ _ _

4. Things that _____ light

    make shadows.

    _____

    _ _ _ _ _ _ _ _ _

5. When you see your _____

    your body has blocked the light.

What are some sources of light?

_____

– – – – – – – – – –

6. The _____ provides most

of the light on Earth.

_____

– – – – – – – – – –

7. Light also comes from _____ .

8. Other lights, like lamps, are made by

_____

– – – – – – – – –

_____ .

9. When light hits an object, it

_____

– – – – – – – – –

_____ off the object into

your eyes.

Critical Thinking

10. How do you use light to stay safe when

crossing a street?

_____

– – – – – – – – – – – – – – – – – –

_____

Light

Write what each picture shows about light and how people use it.

1.

2.

3.

© Macmillan/McGraw-Hill

Light

Fill in the blanks. Use the words from the box.

see	streetlights	Sun

Without light there would be
no life on Earth. The light of the
_____

— — — — — — — — —

_____ helps plants grow
and make food.

People also make some of the light they
_____

— — — — — — — — —

need to _____ things.
Lights like traffic signals and
_____

— — — — — — — — —

_____ help keep us safe.
We depend on light in many different ways.

© Macmillan/McGraw-Hill

Turn On the Lights

✏ **Write About It**

Write a story about the different lights in the picture. Use details. How would the lights help you if you were there?

Getting Ideas

Use the chart below. Remember a cause is why something happens. An effect is what happens. Think about what happens when you use light.

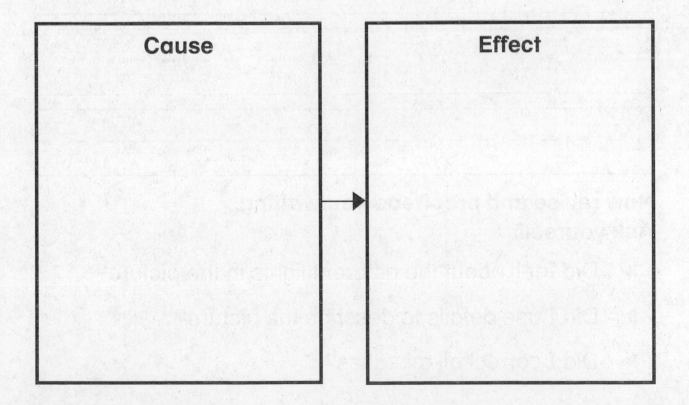

Cause	Effect

Drafting

Write a sentence. Tell a big idea. Why are lights important?

_____

– – – – – – – – – – – – – – – –

_____

– – – – – – – – – – – – – – – –

_____

Now write your story. Begin with the sentence you wrote. Tell about lights in the picture. Use details to describe them. Tell how the lights help you.

_____

– – – – – – – – – – – – – – – –

_____

_____

– – – – – – – – – – – – – – – –

_____

_____

– – – – – – – – – – – – – – – –

_____

Now revise and proofread your writing. Ask yourself:

▶ Did I tell about the different lights in the picture?

▶ Did I use details to describe the picture?

▶ Did I correct all mistakes?

Electricity

Use your book to help you fill in the blanks.

How do you use electricity?

I. We plug lamps and computers into outlets

_____

– – – – – – – – –

because these things need _____

to use them.

_____

– – – – – – – – –

2. Electricity is a form of _____ .

_____

– – – – – – – – – –

3. It gives some things the _____

to work.

_____

– – – – – – – – –

4. Electrical _____ bring

electricity into your school and home.

_____

– – – – – – – – –

5. You can also use _____

to get electricity.

© Macmillan/McGraw-Hill

Name _____

6. Many things we use need electricity to

_____

_ _ _ _ _ _ _ _ _ _ _

_____ .

7. Some electrical things are _____ .

8. You should never use electricity near

_____

_ _ _ _ _ _ _ _ _ _ _

_____ .

Critical Thinking

9. What can happen if electrical wires blow

down in a storm?

_____

_ _ _ _ _ _ _ _ _ _ _ _

_____

_ _ _ _ _ _ _ _ _ _ _

_____

Electricity

Tell how each thing uses electricity or not.

1.

_____
- - - - - - - - - - - - - - -
_____
- - - - - - - - - - - - - - -

2.

_____
- - - - - - - - - - - - - - -
_____
- - - - - - - - - - - - - - -

3.

_____
- - - - - - - - - - - - - - -
_____
- - - - - - - - - - - - - - -

4.

_____
- - - - - - - - - - - - - - -
_____
- - - - - - - - - - - - - - -

Electricity

Fill in the blanks. Use the words from the box.

electricity	energy	heats

Did you have toast for breakfast? If so,

_____

_ _ _ _ _ _ _ _ _ _

the heat _____ used to

make it came from electricity.

When you turn the toaster

_____

_ _ _ _ _ _ _ _ _ _

on, _____ flows

through the wire in the outlet. It

_____

_ _ _ _ _ _ _ _ _ _

_____ the toaster so that

it toasts the bread.

Many other useful machines need

electricity to work.

© Macmillan/McGraw-Hill

Electricity at Home

✏️ ## Write About It

Write a story about how this
family could make dinner without
electricity.

Getting Ideas

Look at the picture. What do you notice?
Write how electricity is being used in the
picture. Then write how the picture would be
different without electricity.

Electricity	Without Electricity
_____	_____
_____	_____
_____	_____
_____	_____
_____	_____
_____	_____
_____	_____
_____	_____

Drafting

Begin your story. Look at the picture. Write a sentence. Tell why electricity is important.

_____

- -

_____

_____

- -

_____

Now write your story. Tell how the picture would be different without electricity. Make sure your story has a beginning, middle, and end.

_____

- -

_____

_____

- -

_____

_____

- -

Now revise and proofread your writing.
Ask yourself:

▶ Did I tell a beginning, middle, and end?

▶ Did I tell how the picture would be different?

▶ Did I correct all mistakes?

Energy Everywhere

Unscramble each word. Write it on the line.

_____
– – – – – – – – –

1. abvreit   When things _____

they cause sounds.

_____
– – – – – – – – –

2. tiyiterclec   People use _____

to light their home.

_____
– – – – – – – – –

3. eaht   We get both _____

and light from the Sun.

4. ynreeg   We use many forms of
_____
– – – – – – – – –

_____ to help us

do work.

5. hicpt   Sounds with a high
_____
– – – – – – – – –

_____ are caused

by fast vibrations.

© Macmillan/McGraw-Hill

Name _____

Circle the word that tells what kind of energy is being used.

6.

light

heat

sound

7.

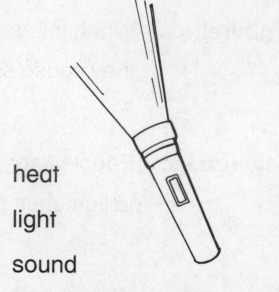

heat

light

sound

8.

electricity

heat

light

9.

light

sound

heat